T0176775

Magnetic Sensors for Biomedical Applications

IEEE Press Series on Sensors

Series Editor: Vladimir Lumelsky, Professor Emeritus, Mechanical Engineering,
University of Wisconsin-Madison

Sensing phenomena and sensing technology is perhaps the most common thread that connects just about all areas of technology, as well as technology with medical and biological sciences. Until the year 2000, IEEE had no journal or transactions or a society or council devoted to the topic of sensors. It is thus no surprise that the IEEE Sensors Journal launched by the newly-minted IEEE Sensors Council in 2000 (with this Series Editor as founding Editor-in-Chief) turned out to be so successful, both in quantity (from 460 to 10,000 pages a year in the span 2001–2016) and quality (today one of the very top in the field). The very existence of the Journal, its owner, IEEE Sensors Council, and its flagship IEEE SENSORS Conference, have stimulated research efforts in the sensing field around the world. The same philosophy that made this happen is brought to bear with the book series.

Magnetic Sensors for Biomedical Applications
Hadi Heidari, Vahid Nabaei

Magnetic Sensors for Biomedical Applications

Hadi Heidari

Vahid Nabaei

IEEE Press Series on Sensors
Vladimir Lumelsky, Series Editor

IEEE PRESS

WILEY

Published by John Wiley & Sons, Inc., Hoboken, New Jersey.
Published simultaneously in Canada.

Limit of Liability/Disclaimer of Warranty
While the publisher and author have used their best efforts in preparing this book, they make no representations or warranties with respect to the accuracy or completeness of the contents of this book and specifically disclaim any implied warranties of merchantability or fitness for a particular purpose. No warranty may be created or extended by sales representatives or written sales materials. The advice and strategies contained herein may not be suitable for your situation. You should consult with a professional where appropriate. Neither the publisher nor author shall be liable for any loss of profit or any other commercial damages, including but not limited to special, incidental, consequential, or other damages.

For general information on our other products and services or for technical support, please contact our Customer Care Department within the United States at (800) 762-2974, outside the United States at (317) 572-3993 or fax (317) 572-4002.

Wiley also publishes its books in a variety of electronic formats. Some content that appears in print may not be available in electronic formats. For more information about Wiley products, visit our web site at www.wiley.com.

Library of Congress Cataloging-in-Publication Data

hardback: 9781119552178

Cover Design: Wiley
Cover Images: © G7 Stock/Shutterstock, © Line-design/Shutterstock,
© Maksim M/Shutterstock

Set in 10/12pt Warnock by SPi Global, Pondicherry, India

Printed in the United States of America

V10015720_112019

Contents

Preface

Magnetic field sensors are widely being used for various applications, ranging from navigation to industrial as well as biomedical applications. A large variety of magnetic sensors based on miniaturized and integrated Hall effect and magnetoresistive (MR) sensors to bulky superconducting quantum interference devices (SQUIDs) have been employed in automotive, information storage, and medical marketplaces. Biomedical magnetic sensors which can capture magnetic fields from biological tissues (e.g. brain, muscle, and cardiac) as well as point-of-care diagnostics were in the center of attention in the past couple of decades. In this regard, highly sensitive, low noise, low power, and integrated magnetic sensors are required to be portable, wearable, and implantable. This book presents the progress and state-of-the-art toward this direction.

As one of the most well-known magnetic sensors, Hall sensors are compact and versatile devices, exhibiting a high magnetic moment sensitivity over a wide field range in both low and room temperature conditions. Moreover, they generally offer a highly linear response, without being affected by magnetic saturation. These sensors have been studied and investigated both numerically and experimentally for different medical applications, such as Hall magnetometry on nanostructures and detection of magnetic beads as a label.

Among the magnetic sensors, it has always been to the benefit of MR sensors that thin-film technology and large magnetoresistance were the interested objects of other often better financially supported activities. For fundamental physics, the investigations of phenomena in thin magnetic films were always valuable because it was easier to explain and understand the magnetism in this specific form of magnetic materials. The most significant support for researches of the magnetoresistance in thin film arises from the success of random access memory technology because it is obvious that the next generation of high-density recording heads will be based on large MR effects.

In this book, the physics of magnetic sensors, their fabrication technologies, and applications are presented. For example, the technology of giant magnetoresistive (GMR) sensors is very sophisticated and refined and therefore it is described as an excellent example of the thin-film technology. Thus, this book not only describes some classes of magnetic sensors but also discusses many more universal subjects, such as magnetism, thin-film technology, fabrication techniques, magnetic measurements, and applications. Therefore, it should be useful for various kinds of readers, especially addressed to engineers. For high-quality specialists, it should be interesting as a comprehensive review of all available knowledge to date. For the sensors' working mechanism, the mathematical descriptions are also presented; therefore, the level is also applicable to students and practising engineers. In this book, there is much computational information about magnetic sensor performance because the authors are engineers who have been active for numerous years in the field of designing and computational modeling of different magnetic sensors and also magnetic measurements.

As one of the most important applications for magnetic sensors, biomedical applications have been addressed and discussed comprehensively. For example, Hall sensors for the detection and even counting of individual magnetic nanobeads, which can be used as labels for medical imaging, drug delivery, and manipulation of biological species, have been discussed. MR sensors for detection of bioanalytes, monitoring of magnetic fluids, biomolecular recognition experiments, and neural activity recording as an ultrasensitive magnetic array have been addressed. We have also presented applications of nuclear magnetic resonance (NMR) sensors for magnetic resonance imaging, electron spin resonance (ESR) sensors for direct detection of paramagnetic species and ESR oximetry, as well as SQUIDs for biomagnetism, magnetoencephalography, and magnetocardiography. That is why this book is entitled "magnetic sensors for biomedical applications."

All parts of the book, Hall effect, MR, NMR, and SQUID sensors and their applications, are written as separate chapters (with independent references and independent logical concept). So, it is not necessary to read the book "from the beginning." If, for example, readers are interested in the one specific sensor, it is sufficient to read only the relevant chapter. In all parts, the structure is the same, consisting of overview, sensor structure and working mechanism theory, sensor different classes, and finally applications. Therefore, it is worth reading this book to gain a better understanding of the features of magnetic sensors and hopefully to use them more effectively. In addition, the authors also recommend reading this book as a story that is one of the most fascinating events of advanced technology. The sensors fabricated in nanometric atomic scale exhibit extraordinary and not fully understood phenomena. The expected future applications of magnetic sensors, especially in medical applications and data storage systems, open new perspectives for the whole of science.

Hadi Heidari and Vahid Nabaei

1

Introduction

1.1 Overview

Magnetic field sensors have been born and used for many years in early applications of direction finding or navigation. Today, magnetic sensors are still a primary means of navigation, but many more uses have evolved. The technology for sensing magnetic fields has also evolved driven by the need for improved sensitivity, smaller size, and compatibility with electronic systems. The output of other sensors, such as temperature, pressure, strain, and light sensors, directly reports the desired parameters, while using magnetic sensors to detect direction, presence, rotation, angle, or electrical currents only indirectly detect these parameters. The output signal of a magnetic sensor requires some signal processing to translate it into the desired parameter value. This makes magnetic sensing a little more challenging to apply in most applications, but it allows for reliable and accurate sensing of parameters that are difficult to sense otherwise. One approach to the development of magnetic sensors is the pursuit of an ideal device that meets the demands and limitations of all the possible applications. Such an ideal sensor must have an ultra-high resolution, a wide bandwidth, very low power consumption, as well as being miniature, low cost, which, all together, does not seem realistic. An optimal magnetic sensing device is that which best fits a set of requirements dictated by a specific application.

This book aims to assist the readers in their search for their optimal magnetic sensing system. From the more common and popular Hall effect sensors up to the nuclear magnetic resonance (NMR)-based magnetometers, each chapter describes a specific type of sensor and provides useful information that is necessary to understand the magnetometer behavior, including theoretical background, noise model, materials, electronics, applications, design, and fabrication techniques. In this chapter, we outline the history of magnetic sensors, natural and technical magnetic fields, magnetic terms and units, magnetic microsensors and their properties and classification, magnetic sensor

Magnetic Sensors for Biomedical Applications, First Edition. Hadi Heidari and Vahid Nabaei.
© 2020 by The Institute of Electrical and Electronics Engineers, Inc.
Published 2020 by John Wiley & Sons, Inc.

terminology, and noise in magnetic sensors. In Chapter 2, we focus on Hall magnetic sensors, and discuss Hall effect, characteristics of Hall effect devices such as geometry and material, horizontal and vertical CMOS Hall devices, and Hall sensor applications. Magnetoresistance (MR) sensors' material and principles, classes, modeling and simulation, design and fabrication technologies, as well as their various biomedical applications have been outlined in Chapter 3. In Chapter 4, we review models, instruments, and biomedical applications of different resonance magnetometers such as NMR, magnetic resonance imaging, and electron spin resonance. Chapter 5 provides an overview of fundamentals, fabrication technologies, and biomagnetism applications of superconducting quantum interference devices. Comparisons of Hall sensors with other galvanomagnetic sensors, NMR with electron spin resonance, and conclusion are presented in Chapter 6.

1.2 History of Magnetism Studies and of Its Use in Magnetic Sensors

The expression "magnet" or "magnetic" originates from the region Magnesia in Thessaly (Greece) where magnetic loadstone (Magnetite, Fe, O) is found as a natural resource [1]. The first reports, in Europe, of the attraction and repulsion forces arising between magnetic loadstones were made by Thales of Miletus around 600 BCE. The expression "sensor" is derived from the Latin "sensus" meaning capable of sensitivity. It is gradually replacing previously used expressions such as "measurement pick-up" and "probe." The directional compass can be regarded as the first magnetic sensor since it reacts to the Earth's magnetic field. Its history stretches back over 4000 years and can be traced to the Chinese who first discovered magnetic loadstone as a natural source of magnetism and used it as a directional aid for orientation [2]. The compass became significantly more important in Europe from about 1200 CE onwards, and in particular, around the time of the great seafaring adventurers and explorers. On his transatlantic voyages, Christopher Columbus observed the behavior of the compass as he sailed westward; famous compass-makers are known to have lived in London and Nurnberg around 1500. In 1820, Oersted discovered that a current-carrying wire deflected a compass needle in its vicinity, and with that the age of electromagnetism had arrived. The first mathematical formula describing the correlation between electric current and magnetism through the deflection of a magnetic needle was Biot–Savart's law. Faraday repeated and extended Oersted's experiments and while doing so discovered the Law of Induction in 1831. The first magnetometer to be constructed was the bifilar magnetometer built in 1831 by Gauss and Weber and the Weber–Bussole compass in 1841, which included a magnetic needle used to measure powerful currents. In 1862,

Maxwell created the common theoretical basis for electromagnetism with laws which were named after him, though the expression "permeability" can be traced to Lord Kelvin. Throughout the history of this topic it is evident that the availability of different types of sensors operating on a magnetic basis is very closely linked to the development and availability of special magnetic materials and to the discovery of new physical and magnetic effects. Some of these effects were very soon exploited to make new magnetic sensors, others were not used until much later, and some are yet to be utilized in sensors. Table 1.1 gives a brief history of the use of magnetic effects in sensors, see also [3].

1.3 Natural and Technical Magnetic Fields and Their Order of Magnitude

Both our natural environment and our technical surroundings provide magnetic fields of many different types and orders of magnitude [1]. Many magnetic sensors detect such fields either directly or indirectly using various principles. Hence, it is essential to take a closer look at the types and magnitudes of the various fields.

1.3.1 Natural Magnetic Fields

1.3.1.1 The Earth's Magnetic Field
The most ubiquitous natural field of all is the magnetic field of the Earth which surrounds us perpetually. It is a dipole field, whose field lines originate at the magnetic poles in the interior of the earth, escaping through the surface and reaching into outer space (Figure 1.1). The Earth's magnetic field is used every day by people with compasses to determine the direction or at altitudes of several hundred kilometers to stabilize satellites; on the Earth's surface it can be employed as a constant reference field within certain ranges and times. The precise determination of the parameters of the Earth's magnetic field, namely its magnitude and direction, was one of the great pioneer acts in the field of magnetism and went back to the works of Gauss and Oersted [1].

1.3.1.2 Magnetic Fields in Outer Space
With the aid of satellites and sensitive magnetic-field sensors, magnetic fields in the vicinity of planets and in outer space can also be determined directly, see Table 1.2 and [4].

1.3.1.3 Biomagnetic Fields
Human beings also produce small magnetic fields, which are primarily caused by microcurrents in cardiac, brain, and muscle tissues [5]. They can nowadays

Table 1.1 History in examples: the magnetic effects in first magnetic sensors [1].

Year	Effect	Explanation	Technical use
1842	Joule effect	Change in the shape of ferromagnetic body with magnetization (magnetostriction)	In combination with piezoelectric elements for magnetometers and potentiometers
1846	AE effect	Change in Young's modulus with magnetization	Acoustic delay line components for magnetic field measurement
1847	Matteucci effect	Torsion of a ferromagnetic rod in a longitudinal field changes magnetization	Magnetoelastic sensors
1856	Magnetoresistance (Thomson effect)	Change in resistance with magnetic field	Magnetoresistive sensors
1858	Wiedemann effect	A torsion is produced in a current-carrying ferromagnetic rod when subjected to a longitudinal field	Torque and force measurement
1865	Villari effect	Effect on magnetization by tensile or compressive strength	Magnetoelastic sensors
1879	Hall effect	A current carrying crystal produces a transverse voltage when subjected to a magnetic field vertical to its surface	Magnetogalvanic sensors
1903	Skin effect	Displacement of current from the interior of material to surface layer due to eddy currents	Distance sensors, proximity sensors
1931	Sixtus–Tonks effect	Pulse magnetization by large Barkhausen jumps	Wiegand and pulse-wire sensors
1962	Josephson effect	Tunnel effect between two superconducting materials with an extremely thin separating layer; quantum effect	SQUID magnetometers

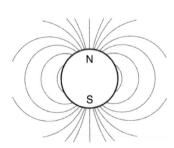

Figure 1.1 Magnetic field formed around the Earth.

Table 1.2 Magnetic field strengths of celestial bodies and objects in outer space [1].

Celestial bodies	Field strength (A/cm)	Flux density (mT)
Galaxies	$\approx 0.15 - 0.25 \times 10^{-6}$	$0.2 - 0.3 \times 10^{-6}$
Mercury (poles)	$\approx 3 \times 10^{-3}$	0.35×10^{-3}
Jupiter (Poles)	Up to 6	Up to 0.8
Mars	0	0
Saturn (equator)	$\approx 2.4 - 16$	$0.3 - 2$
Sun (surface)	$\approx 4 - 8$	$0.5 - 1$
Earth (poles)	≈ 0.5	0.06
A-stars	Up to 28×10^3 A/cm	Up to 3.5 T

be measured with highly sensitive magnetic-field detectors such as flux-gate magnetometers, superconducting quantum interference devices (SQUIDs), and gradiometer coils on the surface of the body, whereas in the past, it had only been possible to measure these currents indirectly by attaching electrodes to the skin and measuring the relevant voltage drops, a technique which forms the basis for the electrocardiogram (ECG). The magnitude of the field strengths and flux densities produced by the heart and brain currents are approximately 50×10^{-3} nT and 1×10^{-3} nT, respectively [6].

1.3.2 Technical Magnetic Fields

1.3.2.1 Magnetic Fields in the Vicinity of Transformers and Electric Motors

Magnetic fields produced in electrotechnical devices, equipment, and plants are usually in the field-strength range of $0.1-10^4$ A/cm [1]. In most cases, these would be AC fields emitted by overhead lines, electrified lines (train or tram lines, etc.), transformers, and electrically powered machinery. Transformers and machinery are mostly operated near the saturation limit of their iron cores, and as such the polarization in the core material attains a level of about 2 T. Similarly, both flux densities in the air gaps of big choke coils or in the gaps between rotors and stators of electric machinery and the stray fields in the vicinity are, as a rule, not much lower.

1.3.2.2 Fields of Permanent Magnets

Except for electromagnets, permanent magnets and magnetic systems are frequently used to produce static magnetic fields, particularly in measurement devices. In such cases, the field is mostly concentrated into a specified volume (e.g. operating air gap) through appropriate design of the magnet and magnetic circuit. The magnitude of the field strengths or flux densities attainable depends on both the remanence and the energy density of the magnetic materials as well as on the geometry of the magnetic circuit. Solenoids with a conventional design, iron-cored coils (electromagnets), and superconducting (solenoid) coils can produce flux densities in the region of 1–100 T depending on their dimensions, their magnitude, and their mode of operation [7–9]. Superconducting coils in medical scanning machines (NMR systems) with a practical diameter of roughly 1 m can produce fields of 1 –2 T, and individual coils with large diameters are used for specified tasks in elementary particle physics and nuclear fusion producing flux densities anywhere from 2 to 10 T.

1.4 Magnetic Terms and Units

The subject of magnetism generates a number of terms and units which are commonly used, the more important ones being included in Table 1.3 [1]. Nowadays, magnetic terms and units are given using the MKSA system, a subsystem of the SI system, and they are based on the four basic units of meter, kilogram, second, and ampere. Although the previously used electromagnetic system of dimensions (CGS electromagnetic units, emu), also known as the Gauss system, is officially no longer acceptable, it is still sometimes found in

Table 1.3 Magnetic terms and units [1].

Term	Quantity	MKSA unit	Subunits	CGS unit	Conversion
Magnetic field strength	H	A/m	1 A/cm = 100 A/m	Oe (Oersted)	1 Oe = 79.58 A/m
Magnetization	M	A/m	See field strength		
Magnetic induction	B	T	$1\,mT = 10^{-3}\,T$	G (Gauss)	$1\,G = 10^{-4}\,T$
Magnetic flux	ϕ	Wb	—	Mx (Maxwell)	$1\,Mx = 10^{-8}\,Wb$
Magnetic polarization	J	T	See induction		
Permeability (absolute)	P	T·m/A	—	G/Oe	
Permeability of vacuum (magnetic constant)	μ_0	$4 \times \pi 10^{-7}$ T·m/A	$0.4 \times \pi 10^{-7}$ T·m/A	1	—

the literature and so a conversion table (Table 1.3) for the CGS units has been included [10].

The most important basic equations are:

$$B = \mu_0 \cdot (H + M) = \mu_0 \cdot H + J \tag{1.1}$$

$$M = \frac{B}{\mu_0} - H \tag{1.2}$$

$$\mu = \mu_0 \cdot \mu_r \qquad \mu_r = \text{Relative permeability} \tag{1.3}$$

$$B = \mu_0 \cdot \mu_r \cdot H \tag{1.4}$$

where, B is magnetic induction, μ_0 is permeability of vacuum, H is magnetic field strength, J is magnetic polarization.

1.5 Magnetic (Micro) Sensors

1.5.1 Definition of Magnetic Sensors

Magnetic materials, i.e. soft and hard magnetic materials and all other materials, which are sensitive to magnetic fields, play a principal role in the nature and operation of magnetic sensors. But at this stage it is not relevant to consider whether metals, metal oxides, or semiconductors are concerned, or which magnetic field is influencing physical properties [11–14]. In textbooks on measurements and control, books on sensors and review articles, sensors have been classified using the following methods [15, 16]: Types of sensors; Physical principles; Properties measured; Sensor applications; Sensor technologies. For example, in [1], a classification system has been selected which is primarily based on physical principles and effects, as follows: Magnetogalvanic sensors; Magnetoelastic sensors; Magnetic-field sensors: saturation-core magnetometers (flux-gate magnetometers) and induction-coil and search-coil magnetometers; Inductive sensors (including eddy-current sensors); Wiegand and pulse-wire sensors; Magnetoresistive sensors; SQUID sensors. Microsensors for a magnetic field are modulating transducers (all semiconductor sensors and all silicon sensors) [17]. They convert the magnetic field, whether it is constant or variable, or even of biological origin, if possible, with a maximal degree of accuracy and reliability, into an output electrical signal (current, voltage, or frequency) with high fidelity. The output energy of the modulating sensors is fed by an external power source through an additional input. More specifically, these microdevices are fabricated using standard semiconductor IC technologies (most frequently, silicon processing technologies). Contemporary microsystems for magnetic fields (hybrid or monolithic) must integrate at least two functions. One of them must be the sensing by an input transducer or sensor of the strength and the direction of this physical measurand; the other can be signal processing (when necessary, including a processor and the corresponding software) and/or an actuator. Elements that locally enhance, compensate, or change the direction of the external magnetic field such as ferrite flux concentrators or coils, can also be installed in the microsystem [1, 18–20]. By using appropriate packages with small dimensions, these MEMS offer a variety of contactless sensing.

1.5.2 Soft and Hard Magnetic Materials for Sensors

Almost all magnetic sensors include magnetic materials in the form of active or passive components, and to a large extent, they determine the concept, construction, and ultimately the sensitivity of the sensors. In view of this, a brief survey of magnetic materials and their most important characteristic is presented in Tables 1.4 and 1.5. These materials are classified according to the IEC system for soft and hard magnetic materials. For textbooks on materials refer to [21, 24]. The magnetic materials listed in Tables 1.4 and 1.5 are closely related to the main classes of magnetic sensors shown in Table 1.6 although some special materials have been added.

1.5.2.1 Shape of the Hysteresis Loop
The most typical characteristic of soft magnetic materials is its hysteresis loop. The shape of the loop can vary greatly and is determined by the type of material and its structure which can be changed by processing and annealing. This is

Table 1.4 Soft magnetic materials [1].

Group	Code	
	A	Irons
	B	Low carbon mild steel
	C	Silicon steel, mainly with 3% Si
Crystal metals	D	Other steels
	E	Nickel–iron alloys (five groups E1 – E5 with 30% – 83% Ni)
	F	Iron–cobalt alloys (three groups F1 – F3 with 23% – 50% Co)
	G	Other alloys as AlSiFe-alloys
Oxides	H	Soft ferrites as NiZn and MnZn oxides
Amorphous metal	I	Amorphous alloys (Fe-based and Co-based alloys)
Powder composite metals	—	Based on Fe and iron alloy powders

Table 1.5 Hard magnetic materials [21–23].

Group	Code	
Crystal metals	R1	Alloys of AlNiCo-type
	R2	Platinum–cobalt alloys
	R3	Iron–cobalt–vanadium (Chromium) alloys
	R6	Chromium–iron–cobalt alloys
	R5	Rare earth cobalt alloys
	R7	Rare earth iron alloys
Amorphous metal	—	Rare earth iron alloys
Oxides	Sl	Hard ferrites as Ba- and Sr-ferrites
	T	Other hard magnetic materials, e.g. magnetically semi hard metals

Table 1.6 Magnetic material for sensors (materials are defined in Tables 1.4 and 1.5).

Sensor class	Magnetically soft Material	Magnetically soft Useful for	Magnetically hard Material	Magnetically hard Useful for
Magnetogalvanic	C, E	Slotted cores yokes	R1, R5, R7	Magnetic circuit
Magnetoelastic	D, E1, I, C, E1	Shafts, surface layers for shafts, laminated core packages, pot-cores	—	—
Fluxgate	E1, H, I	Strips and rods, toroidal cores	—	—
Inductive, eddy current	C, E2, E3, H	Rods, yokes, laminated cores, pot core, rods	R1, R5, R7, S1	Parts
Wiegand, pulse-wire	F, special alloys	Wires	R6	Magnets as rods for switching
Magnetoresistive	E1, I, NiCo, NiFeCo	Resistors	R2, R5, CoCr	Permagnetizing layers

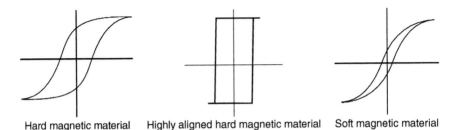

Hard magnetic material   Highly aligned hard magnetic material   Soft magnetic material

Figure 1.2 Three main types of the shapes of hysteresis loops in different materials. *Source:* adapted from [1].

valid for both metals and oxides. The three main types of loop are shown in Figure 1.2 [25].

As the number of magnetic materials currently available is vast, there seems little point in providing extensive tables of material data. This information is published in standard monographs and company catalogues. Instead, we have chosen to present the ranges of variation of the most critical magnetic material terms found in survey diagrams.

1.5.2.2 Saturation Polarization J_s and Coercivity H_c

Two of the most essential material terms are saturation polarization J_s, which gives an idea of the amount of material required for a particular component and coercivity H_c, which provides the basis for classification of the magnetic material concerning its hard and soft magnetic qualities. Figure 1.3 shows a plot of saturation polarization J_s versus the coercivity H_c, where J_s is plotted on a linear scale and H_c on a logarithmic scale, since the materials currently available range from extremely soft alloys, such as permalloy, to commercial iron and steel, covering approximately four to five orders of magnitude.

1.5.2.3 Initial Permeability μ_i

Another critical property of soft magnetic materials is the (relative) initial permeability μ_i. Crystalline 72–83% NiFe alloys and cobalt-based amorphous alloys have the highest initial permeability, with a value of more than 100 000. Mid-range permeabilities in the region of 2000–15 000 occur in NiFe alloys with 36–50% Ni and also in ferrites. Low initial permeability values of 300–1500 are typical for iron, silicon–iron, cobalt–iron, and some ferrites. Even lower values, in the region of 50–300, occur with powder composite materials and ferrites and very low permeabilities (down to values of 5–10) are typical for some special powder composite materials and special ferrites.

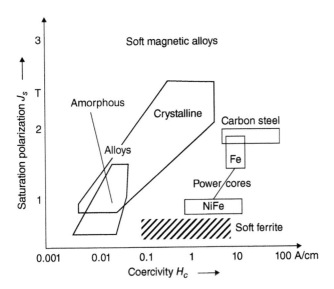

Figure 1.3 Survey on soft magnetic materials plot of saturation polarisation J_s versus the coercivity H_c, where J_s is plotted on a linear scale and H_c on a logarithmic scale. *Source:* adapted from [1].

1.5.2.4 Specific Electrical Resistivity ρ

The difference between the specific electrical resistivity of metals and that of ferrites is several orders of magnitude, as a result of which, for AC-field applications, metals require laminated cores whereas ferrites and powder composites can be used directly as compact cores and parts. Figure 1.4 shows the range of the specific electrical resistivity ρ for metals, ferrites, and powder composites.

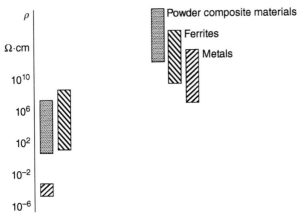

Figure 1.4 Specific electrical resistivity of magnetic materials. *Source:* adapted from [1].

1.5.3   Mechanical Properties of Magnetic Materials

In certain sensor types, the mechanical properties of materials are also important, for example, in magnetoelastic sensors the hardness, yield strength, compressive strength, Young's modulus, etc. must all be considered. Several of the mechanical properties of soft and hard magnetic materials have been compiled in Table 1.7.

1.5.4   Relations Between Sensing Techniques and Sensor Applications

Two categories of applications of magnetic sensors and microsystems can be distinguished: direct and indirect [1, 17, 18, 20, 26]. Direct registrations are those in which only information about the magnetic field (strength and

Table 1.7 Mechanical properties of magnetic materials [1].

Material group		Vickers hardness (HV)	Yield strength (N/mm^2)	Young's modulus (kN/mm^2)
Magnetically soft	Metals Crystalline	80 – 200	150 – 500	100 – 230
	Metals amorphous	800 – 1000	1500 – 2000	~150
			Compressive strength (N/mm^2)	
	Soft ferrites	800	75 – 100	150
Magnetically hard	Powder composite metals	—	60 – 250	~50
			Yield strength (N/mm^2)	
	Malleable alloys like FeCoCr, FeCoV	450 – 950	1000 – 1500	245 – 255
			Compressive strength (N/mm^2)	
	Metals, cat AlNiCo, magnets	400 – 700	120	—
	RE magnets (SmCo, NdFeB)	550 – 650	850 – 1000	110 – 150
	Hard ferrites	600 – 1000	500 – 800	150

direction) itself is required. Indirect sensor systems use the field as an interme-
diary carrier to measure a nonmagnetic quantity (tandem transducers). Examples of direct applications include the following:

- Readout of information stored on disk, tape, and bubble memory
- Recognition of magnetic patterns on banknotes and credit cards
- Magnetometry: control of a magnetic apparatus such as classical and super-
  conducting electromagnets; instrumentation for particle accelerators as well
  as the determination of the full magnetic field vector, its direction, and its gra-
  dient by detecting two or all three vector components
- Magnetic levitation control
- Earth magnetic-field measurement and the electronic compass
- Geomagnetic remote sensing for geological and volcanic surveying
- Attitude control for satellites
- Positioning of aircrafts, ships, missiles, projectiles, or submarines by the perturbations they cause in the geomagnetic field, and for the development of
  global navigation systems
- Biomagnetometry: obtaining diagnostic data by cardiomagnetism, myomag-
  netism, and neuromagnetism to map the functions of the heart, muscles,
  nerves, and brains of humans and animals

Indirect applications are much more common [1, 17, 18, 20, 26–30]. Examples include the following:

- Distance (linear and angular), velocity, speed, and vibration measurements
- Position detection
- Rotation and direction of rotation, e.g. for tachometry
- Collector-less DC motor control
- Keyboards and proximity switches
- Microphones
- Angular displacement detection and angle decoders and synchro resolvers
- Linear and rotary potentiometers, and crankshaft position transducers in
  automobile ignition control
- Automotive antiskid breaking systems
- Nondestructive magnetic methods, including characterization of materials
  and metal detection
- Electrical current and power measurements (watt-hour meters) that do not
  interrupt the current-carrying conductor
- Analog multiplication
- Galvanic isolation
- Traffic detection when a ferromagnetic body is passing
- Measurement of mechanical and chemical parameters, pressure, mass filters,
  and so on, using suitable magneto-modulating systems that contain perma-
  nent magnets.

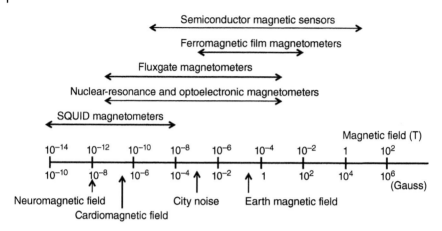

Figure 1.5 Field range of five principally different types of magnetic transducers and microsystems. *Source:* adapted from [17].

These incomplete lists of the magnetic sensor and microsystem applications should suffice to show how universal they are. From the viewpoint of the achieved accuracy, these are some of the most precise instruments available in control-measurement technology. The range of operation of various magnetic sensors and microsystems is systematized in Figure 1.5. The applications selected above clearly show the wide variety of fields met by magnetic (micro) sensors. It ranges from the biosignals, which are lower than 10 fT; passing through the variations of the geomagnetic field (~0.05 mT) through the most important large-scale applications, using permanent magnets with induction fields at about 50 mT. With the recording of magnetic fields ranging from 0.01 to 10 mT, reach to flux densities of several Tesla in the apparatus for nuclear physics and the colossal inductions of millions of Tesla in the new stars and black holes of the universe. Therefore, there is a dynamic range of fields not less than 15–16 orders of magnitude! As seen from Figure 1.5, at least five principally different types of devices are capable of measuring this unique range of magnetic fields: superconductor magnetic sensors such as SQUIDs, nuclear-resonance and optoelectronic magnetic transducers, flux gate magnetometers, ferromagnetic thin-film magnetoresistors, and semiconductor magnetic devices. That is why this broad range cannot be covered by only one type of magnetic sensor microsystem.

1.5.5 Classification of Magnetic Sensors

The great variety of applications, the peculiarities of transducer action and reliability, the growing research activity and the necessity to compare the results, the demanding requirements concerning the quality of used materials, and the

Table 1.8 Classification of the principal figures of merit of magnetic sensors [17].

$OUT(B)_{C = const}$	$OUT(C)_{B = const}$	System descriptors
Sensitivity	Noise	Excitation
Nonlinearity	Offset	Input impedance
Calibration	Cross-sensitivity	Output impedance
Range	Drift	Size
Frequency response	Creep	Weight
Directivity	Temperature error	Packaging
Resolution	Operating life	Sensor material
Accuracy	Reliability	Environmental condition
Hysteresis	Long-term stability	Device design
Error (reversibility error)	Response time	
Output form		

tendency for unification necessitate the development of clear criteria for evaluation of the different sensors, including the magnetic sensor microsystems. Consideration of all figures of merit can be made more comprehensible if magnetic sensors are classified into three groups [20], as follows:

1) The change of the output (*OUT*) in a magnetic field B, keeping (or leaving) constant other possible external influences, e.g. temperature T, pressure P, and radiation Φ, summarized here with the symbol C. The denotation of these characteristics is $OUT(B)_C$.
2) The behavior of the output signal *OUT* as a function of all possible external influences C at a constant (or absent) magnetic field B, denoted $OUT(C)_B$.
3) The parameters describing the magnetic sensor as a device, denoted SD (system descriptors).

In Table 1.8, the principal figures of merit of magnetic-field microsensors are presented according to this classification.

1.6 Characteristics of Magnetic Sensors

1.6.1 Characteristics Related to $OUT(B)_C$

1.6.1.1 Magnetosensitivity
By definition, the sensitivity S or the transduction efficiency is the ratio of the change of the output signal (current, voltage, or frequency) to the variation of the external magnetic field B at a constant T, P, Φ, etc. Both an absolute

sensitivity and a relative sensitivity of the modulating magnetosensors can be defined [1, 20, 27, 31]. Equations (1.5)–(1.7) define the absolute sensitivity when the output is a current I, voltage V, or frequency f, respectively:

$$S_A^{(V)} = \left|\frac{\partial V}{\partial B}\right|_C, \ [\mathrm{VT^{-1}}] \tag{1.5}$$

$$S_A^{(I)} = \left|\frac{\partial I}{\partial B}\right|_C, \ [\mathrm{AT^{-1}}] \tag{1.6}$$

$$S_A^{(f)} = \left|\frac{\partial f}{\partial B}\right|_C, \ [\mathrm{HzT^{-1}}] \tag{1.7}$$

The relative magnetosensitivity is determined by the ratio of the absolute sensitivity to the supply current or voltage applied to the additional transducer input. The figure of merit is the current-related sensitivity S_{RI} (Eq. 1.8) when a supply current I_S feeds the additional input, and the voltage-related sensitivity S_{RV} (Eq. 1.9) when a supply voltage V_S feeds the additional input:

$$S_{RI}^{(V)} = \frac{S_A^{(V)}}{I_S} = \left|\frac{1}{I_S}\frac{\partial V}{\partial B}\right|_C, \ [\mathrm{VA^{-1}T^{-1}}]$$

$$S_{RI}^{(I)} = \frac{S_A^{(I)}}{I_S} = \left|\frac{1}{I_S}\frac{\partial I}{\partial B}\right|_C, \ [\mathrm{T^{-1}}]$$

$$S_{RI}^{(f)} = \frac{S_A^{(f)}}{I_S} = \left|\frac{1}{I_S}\frac{\partial f}{\partial B}\right|_C, \ [\mathrm{HzA^{-1}T^{-1}}] \tag{1.8}$$

$$S_{RV}^{(V)} = \frac{S_A^{(V)}}{I_S} = \left|\frac{1}{V_S}\frac{\partial V}{\partial B}\right|_C, \ [\mathrm{T^{-1}}]$$

$$S_{RV}^{(I)} = \frac{S_A^{(I)}}{V_S} = \left|\frac{1}{V_S}\frac{\partial I}{\partial B}\right|_C, \ [\mathrm{AV^{-1}T^{-1}}]$$

$$S_{RV}^{(f)} = \frac{S_A^{(f)}}{V_S} = \left|\frac{1}{V_S}\frac{\partial f}{\partial B}\right|_C, \ [\mathrm{HzV^{-1}T^{-1}}] \tag{1.9}$$

The relative sensitivity should be suitable to the comparative analysis of magnetic microdevices, while in the purely practical applications, the absolute sensitivity is preferred.

1.6.1.2 Nonlinearity
When the ideal output characteristic of the magnetosensor is a straight line, the deviation of the real output characteristic from it is a relative error that is termed a nonlinearity (NL). This parameter is determined by the expression

$$NL \equiv \frac{|OUT_1 - OUT_2|}{(OUT_2)} 100\% \tag{1.10}$$

where OUT_1 is the measured value of the output signal at a given fixed magnetic induction B_o, and OUT_2 is the corresponding value from the straight line at a field B_o. This line is the best fit to the measured output values.

1.6.1.3 Calibration
This figure of merit is necessary to determine the real output signal $OUT(B)$ of any magnetosensor. The calibration is a test during which known values of the magnetic induction (taking into account the sign of B) are applied to the micro-transducer and the corresponding output readings are recorded [20, 31].

1.6.1.4 Sensor Excitation
An important feature of the magnetosensitive device is the manner of their excitation by the field B, i.e. whether the vector B is perpendicular or parallel to the active surface of the structure [20, 31]. Depending on the orientation of B with respect to the plane of the device, we have divided the magnetic microsensors known so far into orthogonal and parallel-field transducers.

1.6.1.5 Frequency Response
The dynamic behavior of the magnetic field sensors is of primary importance in the measurement of fast variations of the field B. The frequency response is the dependence of the amplitude ratio of the output signal to the AC field B on the frequency of the external sine wave of the AC field B within a specified frequency interval. It is generally accepted that the dynamic degradation of a microtransducer in an AC field B starts with a reduction of the output by a factor $1/2$, i.e. by 3 db [20, 31].

1.6.1.6 Resolution
This characteristic determines the smallest possible step change of the magnetic induction, which can be detected on the sensor output [20, 31].

1.6.1.7 Error
This figure of merit is the algebraic difference between the magnetic field recorded at the sensor output and its exact value. The error is expressed as a percentage of the full-scale output. The error curve is a graphical representation of the errors obtained from a certain number of calibration cycles. There is a slight error given in percent termed reversibility error. It expresses the difference in the output readings of the magnetotransducer at fixed value B_o with changes in the magnetic-field B direction (from positive to negative and vice versa). This error strongly depends on the difference in the contact area and nonuniformity of the sensor material used [18, 20, 31].

1.6.1.8 Accuracy
This characteristic is defined as a ratio of the error to the full output scale expressed as a percentage. The accuracy can be presented in the same units by which the magnetic field is measured as within ± ... % of the full scale output [18, 20, 31].

1.6.1.9 Hysteresis
This parameter determines the maximal change in the output signals for any fixed value B_o of the magnetic induction within the specified range. The value B_o is reached first by increasing and then by decreasing the external field B. The hysteresis is expressed as a percentage of the full-scale output during any calibration cycle [18, 20, 31].

1.6.1.10 Repeatability
The ability of a magnetic microsensor to give the same output when the same measurand value B_o is applied. It is expressed as a percentage of the full-scale output. Many calibration cycles are needed to determine the repeatability [18, 20, 31].

1.6.2 Characteristics Related to $OUT(C)_B$

1.6.2.1 Noise
The output electrical noise is a fundamental property that determines the lowest detected value of the magnetic field B_{min}[1, 18–20, 27, 31]. In this output signal (voltage or current) with a random amplitude and random frequency, which has nothing in common with the measurand B, the most disturbing component is $1/f$ noise. By selection of a more qualitative semiconductor material and more sophisticated technological steps, the $1/f$ noise can be significantly reduced. Often this figure of merit is presented as an equivalent magnetic-field noise at a signal-to-noise ratio equal to one.

1.6.2.2 Offset
The offset is a parasitic output signal in magnetic microsensors of a modulating type (in most cases with a differential output) in the absence of a magnetic field B. Without any supplementary information about the value of induction B, the offset cannot be distinguished from the useful output signal (voltage, current, or frequency):

$$OUT(B) = SB + OUT'(B = 0) \qquad (1.11)$$

where $OUT'(B = 0)$ is the offset that is most frequently the result of a structural or electrical asymmetry of the sensor. This error is often expressed as within + ... % of the full-scale output [1, 18–20, 27, 31]. A more relevant definition of the offset is its expression as a signal of an ideal sensor (without offset) generated by an equivalent magnetic induction $B_{off,\,eq} \equiv B_{off}$:

$$B_{off,eq} \equiv \frac{V_{off}}{S_A^{(V)}}; \quad B_{off,eq} \equiv \frac{I_{off}}{S_A^{(I)}}; \quad B_{off,eq} \equiv \frac{f_{off}}{S_A^{(f)}} \qquad (1.12)$$

1.6.2.3 Cross-Sensitivity and Temperature Error

The influence of one or more measurands as pressure P, light Φ, temperature T, and so forth, on the magnetosensitivity is an undesired parasitic output signal termed cross-sensitivity (CS) [1, 18–20, 27, 31]. The figure-of-merit CS is expressed as $CS \equiv (1/S\ \partial S/\partial C)$, where S is the sensor magnetosensitivity (Eqs. 1.5–1.9) and C is the source of perturbation. Most frequently this is the temperature T; consequently,

$$TC \equiv \left(\frac{1}{S}\right)\left(\frac{\partial S}{\partial T}\right)100\% \qquad (1.13)$$

In this case, TC is the temperature coefficient of magnetosensitivity, measured as % K^{-1} or % C^{-1}. The temperature interval $(T_{min},\ T_{max})$ within which the value of the TC is constant should be specified.

1.6.2.4 Drift and Creep

The drift is an undesirable slow change in time of the output signal at a constant magnetic field B. The drift is in no way connected with the measurand B. Drift can be caused by temperature, pressure, light, and so on. The creep, which is also a parasitic magnetosensor fluctuation, is a weak and continuous change at the output at constant magnetic field B and all other environmental parameters such as T, P, and Φ [20, 31].

1.6.2.5 Response Time

This figure of merit determines the time it takes for the sensor output signal to reach its final value as a result of a step change of the magnetic field. This parameter is indicated in the handbooks, for example, as "95% response time ... μ_s" [18, 20, 31].

1.6.3 Characteristics Related to the System Description

1.6.3.1 Electrical Excitation

This is the external electrical voltage and/or current applied to a magnetic modulating microsensor for its proper operation.

1.6.3.2 Input and Output Impedance

The input impedance presented to the power supply is measured between the excitation terminals. The output impedance is measured between the output leads of the magnetic sensor under conditions with open-circuited additional input terminals.

1.6.3.3 Environmental Conditions

Magnetic-field microsensors most frequently function under the following environmental conditions: temperature (25 ± 10) °C or (77 ± 18) °F, relative humidity 90% or less, and barometric pressure 26–32 inches Hg. The remaining figures of merit of the magnetic devices, as well as other characteristics, methods, and circuits for their determination, can be found in the literature [18, 20, 31] and the references therein.

1.7 Magnetic Noise

Any macroscopic quantity in a system such as voltage, current, or resistance is subject to fluctuations around its mean value. These fluctuations are created by the random contributions to the transport or by internal displacements of atoms [32]. The first point of view is to consider that these fluctuations are strongly related to the properties of the material and their study gives a new approach to understand processes in condensed matter. A second point of view more related to applications is to call these fluctuations noise which is related to the quality of the material investigated. Reduction of this noise becomes a target for magnetic sensor applications.

1.7.1 Noise Formalism

In this part, we give some bases necessary for the analysis of noise measurements in GMRs with a highlight on some common traps in noise treatment. We will not address here extensively the quantum approach of fluctuations as magnetoresistive sensors have size and working temperature, which can be handled with a classical treatment.

1.7.1.1 Fluctuations, Average and Distribution

We consider a fluctuating quantity $V(t)$. The voltage V is the most common quantity measured in magnetic sensors, but the following treatment can be applied to any relevant quantity like the current, the resistance, or the charge. In an experiment, $V(t)$ is often measured at discrete times t_1, ..., t_n by some acquisition system but it can also be treated in an analog way with some integration, derivation, analog multiplication, or other mathematical operation.

Any nonlinear operation has to be carefully handled in order to avoid spurious deformation of the signals. Then a number of new quantities can be derived from that measurement. The first one is the average

$$\bar{V} = \frac{1}{n}\sum_{i=1}^{n} V(t_i) \tag{1.14}$$

A formal definition of the average, easier to manipulate is:

$$\bar{V} = \lim_{T\to\infty} \frac{1}{T}\int_0^T V(t)dt \tag{1.15}$$

where T is the duration of the measurement. In the following, we use the second formalism. It should be noticed that this average is also a fluctuating quantity on a time scale T corresponding to the total acquisition time. In common systems, if T is long enough this average is representative of V but in some cases like in presence of slow magnetic relaxation, $\bar{V}$ can be very dependent of the history of the system. The second quantity is the variance of V. It is defined as

$$\sigma^2 = \frac{1}{n}\sum_{i=1}^{n} (V(t_i) - \bar{V})^2$$

or

$$\sigma^2 = \lim_{T\to\infty} \frac{1}{T}\int_0^T (V(t_i) - \bar{V})^2 dt \tag{1.16}$$

It measures the exploration in amplitude of V. σ is an easy comparison of different signals but sometimes a spurious frequency may appear in the signal like the line frequency (50 or 60 Hz) which dominates and sets the value of σ. For large systems, the knowledge of σ and $\bar{V}$ is sufficient to know the distribution function ρ of V. This is due to a very useful theorem, the central limit theorem, which demonstrates that a sum of n random identical quantities tends very rapidly to a normal (or Gaussian) distribution law when n grows.

$$\rho(V) = \frac{1}{\sqrt{2\pi\sigma^2}}\exp\left(\frac{(V-\bar{V})^2}{2\sigma^2}\right) \tag{1.17}$$

For example, the voltage across an MR with a fixed sensing current can be divided in smaller identical MRs which individually fluctuates and hence one can easily demonstrate that the distribution of the voltage follows a normal distribution. Another very common fluctuation is the random jump between two discrete levels. This type of fluctuation is, for example, the Random Telegraph Noise (RTN) described later. The jumps are arising from a level 1 to a level 2 with a w_1 probability to stay at level 1 and w_2 probability to stay at level 2.

The mean period of these jumps is τ. The distribution is then showing peaks centered on the values of the levels with a width related to the additional noise. Hence, analysis of the distribution may help to distinguish the presence of discrete levels.

1.7.1.2 Correlations

In time domain, correlations are very important because they give a measure of the similarities between two functions (cross-correlations) or for the same function but at two different times (autocorrelation). The autocorrelation is a measure of the memory of the system. The general form of the correlation of two functions X and Y is defined as the average of the functions at a time difference of τ.

$$C_{xy}(\tau) = \lim_{T \to \infty} \frac{1}{T} \frac{1}{2\pi} \int_0^T X(t) Y^*(t-\tau) dt \tag{1.18}$$

The autocorrelation function is then

$$g_x(\tau) = \lim_{T \to \infty} \frac{1}{T} \frac{1}{2\pi} \int_0^T X(t) Y^*(t-\tau) dt \tag{1.19}$$

The autocorrelation at $\tau = 0$ is simply the mean square value of the fluctuations. For example, a thermal noise has an autocorrelation function which is zero except for $\tau = 0$. This is the signature of a totally random process. However, if you measure a thermal noise through a filter, you may find a nonzero autocorrelation due to the memory injected by the filtering.

It should be noticed that the experimental correlations are dependent on the initial time chosen for the acquisition. Hence, there is a strong hypothesis behind, usually fulfilled by GMR and TMR systems, the stationarity of the system which assumes that the system is at equilibrium and these quantities are independent on the initial measurement time.

1.7.1.3 Frequency Space and Spectral Density

Amplitude and amplitude distribution of fluctuations are analyzed by the tools described previously but spectral analysis gives the frequency dependence of the noise which is essential for separating and understanding the noise sources. The way to switch from time domain to frequency domain is the Fourier transform.

$$V(\omega) = \frac{1}{2\pi} \int_{-\infty}^{\infty} V(t) e^{i\omega t} dt \tag{1.20}$$

We also introduce a more experimental Fourier transform which considers that $V(t)$ is zero outside of the measuring time.

$$V_T(\omega) = \frac{1}{2\pi} \int_{-\infty}^{\infty} V(t) e^{i\omega t} dt \tag{1.21}$$

This well-known form of Fourier transform is however a source of lot of errors in experiments. Firstly, if the total acquisition time is T, the lowest achievable frequency is $2\pi/T$ and the highest is given by $2\pi/t_{acq}$ where t_{acq} is the acquisition time interval. Secondly, if signals are present in the fluctuations at higher frequencies, they will appear by folding in the frequency range. This is the reason why low pass filtering is necessary for noise measurements. Thirdly, Fast Fourier Transform (FFT) algorithms are generally used due to the gain in time of data treatment. As the data are finite in time, a window function is applied on the data. The simplest one is just a rectangular window. The data are then considered as 0 outside the window and multiplied by 1 in the window. However, the jump at the edges created spurious oscillations in the Fourier transform. More sophisticated windows are hence used. The most common are Hann, Hamming, and Gaussian windows. The Hann function consists of multiplying the data by a cosine function, which vanishes for the first and last points. So the quantity calculated becomes:

$$V_{Hann}(\omega) = \frac{1}{2\pi} \int_0^T V(t)e^{i\omega t} \left(1 - \left|\cos\left(\frac{\pi t}{T}\right)\right|\right) dt \qquad (1.22)$$

Or in the case of Gaussian window,

$$V_{Gauss}(\omega) = \frac{1}{2\pi} \int_0^T V(t)e^{i\omega t} e^{-\frac{(t-(T/2))^2}{2(\sigma T/2)^2}} dt \qquad (1.23)$$

On rather flat signals like noise, the windowing has no impact but in case of a strong line signal, it allows suppressing artefacts. The total energy of the signal is given by

$$E = \int_0^T |V(t)|^2 dt \qquad (1.24)$$

which can be expressed through the Parseval theorem by

$$E = 2\pi \int_{-\infty}^{\infty} |V(\omega)|^2 d\omega \qquad (1.25)$$

Hence the average power associated to the fluctuations can be defined by:

$$P = \lim_{T\to\infty} \frac{1}{T} \int_0^T |V(t)|^2 dt = \lim_{T\to\infty} 2\pi \int_{-\infty}^{\infty} \frac{|V(\omega)|^2}{T} d\omega \qquad (1.26)$$

This allows us to define the power spectral density (PSD) as:

$$S_V(\omega) = \lim_{T\to\infty} 2\pi \frac{|V(\omega)|^2}{T} \qquad (1.27)$$

The PSD is given for voltage fluctuations in V^2/Hz. A very important point is the relation between the PSD and the autocorrelation function which is the Wiener–Khintchine theorem:

$$S_V(\omega) = 2\int_0^\infty g(\tau)\cos(\tau\omega)|V(\omega)|^2 d\tau \qquad (1.28)$$

A major example is an exponentially decreasing autocorrelation function. It is the case for a relaxation process but also for the current in a simple R–L circuit. The autocorrelation decreases with a characteristic time τ_C and is written as:

$$g(\tau) = g(0)\exp\left(-\frac{\tau}{\tau_C}\right) \qquad (1.29)$$

The corresponding PSD is

$$S_V(\omega) = 4g(0)\frac{\tau_C}{1+\omega^2\,\tau_C^2} \qquad (1.30)$$

This spectrum called Debye–Lorentzian spectrum presents a flat response at low frequencies and a $1/f^2$ decrease at high frequencies with a corner at $1/\tau_C$. This exponential decrease of the autocorrelation function is also valid for RTN noise and hence the RTN signature in the PSD is this Debye–Lorentzian spectrum. Another important remark is related to very slow decrease of the DC level in noise measurements determining the formal spectral density from a noise mechanism. Very often noise measurements setups present a very low frequency high pass filter (0.1 Hz or lower) to avoid amplifier saturation by DC offsets. If there is an external perturbation like a short pulse on the DC line or a jump in the DC signal, this induces a decrease of the output signal with a very long characteristic time and hence a Debye–Lorentzian spectrum with a corner frequency below the measurement range. For that reason, in case of measurement of a $1/f^2$ decrease at very low frequencies, investigation of the DC level fluctuations should be done.

1.7.2 Sensitivity, Signal-to-Noise Ratio, and Detectivity

Noise PSD is given in V^2/Hz but it is usually more convenient to compare a signal given in V to the square root of the PSD which is in $V/\sqrt{Hz}$. In order to evaluate a signal-to-noise ratio, a reference signal at a known frequency generated by a coil is often used. If a signal $V_0\cos(\omega t)$ is seen in the acquisition system, its PSD is the power associated to this signal taken on a bandwidth of 1 Hz so 1 s and it corresponds to $V_0^2/2$. This allows direct calibration of a sensor.

The sensitivity for a magnetoresistive signal, β, is usually given in V/V/T. Typical values for GMR sensors are 20–40 V/V/T or 2–4%/mT.

The output voltage of a magnetoresistive sensor can be written as:

$$V_{out} = ((R_0) + \delta R(H))I \tag{1.31}$$

If the sensor is linearized and well centered, the output can be written as:

$$V_{out} = \left(R_0 + \frac{\delta R}{\delta H}\cdot H + ...\right)I \tag{1.32}$$

where the terms with higher H power are small $\delta R/\delta H$ is given in Ohms/Tesla. The sensitivity is given by $(\delta R/\delta H)/R_0$.

In order to compare different sensors, it is very convenient to use the field equivalent noise PSD, sometimes called detectivity. It corresponds to the PSD divided by the sensitivity. For example, if a sensor exhibits a thermal noise of 1 nV/$\sqrt{Hz}$ and a sensitivity of 20 V/V/T, the corresponding detectivity will be 50 pT for 1 V of bias voltage.

1.7.3 Different Sources of Noise

1.7.3.1 Separation of Magnetic and Nonmagnetic Noise
It should be noticed that noise in magnetic sensors can or cannot be magnetic-field dependent. A magnetic field-dependent noise appears or disappears with the application of an external field. This is, for example, the case of a part of the 1/f noise in magnetoresistive junctions. One way to separate it is to measure the noise of the sensors in the whole range of operation.

1.7.3.2 Frequency-Independent Noise (Thermal or Johnson–Nyquist Noise), Shot Noise
1.7.3.2.1 Thermal Noise
Frequency-independent noise is called white noise and corresponds to processes without any autocorrelation except at zero time. It should be noticed that noise appears flat in the range of measurement because its correlation characteristic time is faster that the minimal sampling time.

The most important noise is the thermal noise, which is directly related to the resistance of the sensor. The first observation of this noise has been done by Johnson [33] and interpreted by Nyquist [34]. The associated PSD is written as:

$$S_V(\omega) = 4RK_BT \tag{1.33}$$

where $K_B = 1.3806 \times 10^{-23}$ J$\cdot$K^{-1} is the Boltzmann constant. There are several ways to demonstrate this relation, but the base is just to say that the energy available for a dipole of resistance R is $K_BT/2$. This formula is valid for frequencies much lower than $K_BT/\hbar$ where $\hbar$ is the Planck constant. This thermal noise cannot be eliminated or reduced except by changing the resistance or the temperature, but it has the advantage to be independent on the voltage applied on the sensor. It should also be noticed that the impact of this noise on the signal-to-noise ratio is directly related to the working bandwidth. The integrated noise will increase as the square root of the bandwidth.

1.7.3.2.2 Shot Noise

This noise is due to the fact that the electrical current is not continuous due to the discrete nature of the electrons. This noise is detectable only if there is a barrier to cross where the quantum nature of electrons is revealed. In metals, the electrons' inelastic scattering length is very small, typically a few nanometer at room temperature and a metallic sensor could be described as a very large number of individual elements of few nanometer in series and hence the shot noise is divided by the square root of this number [35]. For that reason, shot noise is not present in GMR sensors. In magnetic tunnel junctions, a number of theoretical and experimental works have been recently published. The electrons are following a Poisson law when they pass through a barrier (without quantum or correlation corrections) and hence it is possible to calculate the PSD of the shot noise. At $T = 0$, it is simply given by

$$S_I(\omega) = 2eI \tag{1.34}$$

In mesoscopic systems, deviations of the Poisson statistics are observed and modify Eq. (1.34). A review of noise in mesoscopic systems can be found in [36]. In particular, it has been shown that the statistics can be slightly different from a pure Poisson statistics inducing an enhancement or a reduction of the shot noise [37, 38]. This enhancement or reduction is characterized by the Fano factor F.

1.7.3.2.3 Crossover Between Shot Noise and Thermal Noise

When the temperature is increasing, the shot noise expression must take into account thermal fluctuations. A more general formula should be applied [35]:

$$S_V(\omega) = 2eI \coth\left(\frac{eV}{2K_B T}\right) R^2 \tag{1.35}$$

In the two temperature limits ($T \gg 0$ and $T \to 0$), we find the expression of noise given in Eqs. (1.33) and (1.34), respectively.

1.7.4 Low Frequency Noise

1.7.4.1 1/f Noise

$1/f$ noise is a general term referring to a frequency decreasing noise with a power law frequency $1/f^\beta$ where β is an exponent typically of the order of 1. This noise is observed in nearly all fluctuating systems including biological and geological fluctuations. In GMR and TMRs, this low frequency noise is dominant and is often a drawback in performances of magnetoresistive sensors. The origin of the $1/f$ noise is resistance fluctuations and not voltage fluctuations like the thermal noise. Hence, these fluctuations can only be revealed by applying a current in the sensor. Voss and Clarke have demonstrated in [39] that the variance of Johnson noise exhibits a $1/f$ power spectrum demonstrating the resistance fluctuation nature of the $1/f$ noise. This resistance fluctuation behavior implies that the PSD is varying as V^2 or I^2. This is very important because this allows us discriminating between white noise and low $1/f$ noise. This general law can however be modified if the current induces itself modifications of the resistance or local heating. We have observed, in particular, in small GMR devices an increase of $1/f$ noise much faster than I^2.

GMR sensors have an isotropic dependence of the resistance unlike AMR where the signal is depending on the angle between field and current or Hall sensors where the voltage appears in a direction perpendicular to the current. Hence it is impossible to play on current direction to separate resistance variation and external field variation. This is a major issue with GMRs where spinning techniques used in Hall sensors or flipping of the current used in AMR sensors cannot be applied.

The second important point is the size effect on $1/f$ noise. Typically, PSD of the $1/f$ noise is decreasing as the volume of the sensor increases. This can be understood easily by an averaging effect. If we suppose that $1/f$ fluctuations are coming from small individual sources, you can consider that a resistance is the sum of N small resistances r so the total resistance $R = N \cdot r$ but the fluctuations of R are $\sqrt{N}$ larger than the individual r fluctuations. Hence for a given R, the fluctuations are decreasing as $\sqrt{N}$, i.e. as the square root of the volume. This has been observed in several systems for large enough sizes. At very small sizes an individual fluctuator may dominates and this rule is broken. There is a general phenomenological formula proposed by Hooge [40] which allows us to compare various sensors

$$S_V(\omega) = 2\pi \frac{\gamma_H V^2}{N_C \omega} = \frac{\gamma_H V^2}{N_C f} \tag{1.36}$$

where γ_H is a dimensionless constant. This formula is well adapted to semiconductors where the averaging is more on the number of carriers N_C than on the effective volume. For TMRs, the formula becomes:

$$S_V(\omega) = 2\pi \frac{\alpha V^2}{A \omega} = \frac{\alpha V^2}{A f} \tag{1.37}$$

where A is the active surface of the device and α is then a parameter with the dimension of a surface. This last formula describes rather correctly the evolution of the noise with the size of the sensors. We note that $1/f$ noise can exhibit a nonmagnetic and a magnetic component with sometimes different slopes.

The evolution of the noise as function of magnetic field can help to separate the two contributions. Often, the noise recorded under a strong field which saturates the different layers of a magnetic sensor is mainly nonmagnetic and an additional noise due to magnetic fluctuations in the layers appears in the sensitive regions. As described later, the shape of the sensors has a strong impact on the magnetic $1/f$ noise.

In case of TMRs, the $1/f$ noise depends also whether the junction is in the parallel or antiparallel state. In the parallel state, the number of channels opened through the barrier is larger than in the antiparallel state and hence the noise is smaller. This is a direct consequence of the reduction of the effective size of a junction.

1.7.4.2 Random Telegraph Noise

RTN or "popcorn" noise is the noise arising from the fluctuations of a specific source between two different levels. For magnetoresistive sensors, this noise generally appears in devices with a size small enough to let individual defects becoming dominant. However, it can be sometimes observed in reasonably large GMRs at high currents levels. RTN is, as explained before, a fluctuation between two levels with comparable energies and a barrier height able to give a typical characteristic time in the measurement range. Hence, RTN is very dependent on the temperature, field, and applied bias current.

RTN is difficult to handle and a sensor with RTN noise is in general very difficult to use even if it is theoretically possible to suppress partially this noise by data treatment. That suppression requires an RTN with a low fluctuation frequency and two states well separated. The treatment is then based on the recognition of each transition level from low state to high state and suppression of the step.

1.7.5 High Frequency Noise and Ferromagnetic Resonance

At high frequencies, the noise is usually dominated by thermal or shot noise. However, noise peaks can be detected in magnetoresistive sensors due to the fluctuations of magnetic layers. In the GHz regime, thin magnetic films present ferromagnetic resonances with frequencies dependent on the material and on the shape of the sensors. In small elements, quantization of the spin waves induces a large number of resonances. The noise detected is coming from two different sources. The first one is due to the GMR effect. The free layer is fluctuating with more important amplitude at resonance and if a DC current is sent in the device, a voltage at the resonance frequency appears. The second one is due to thermal and shot noises. This noise is amplified by the quality factor of the resonance and hence this appears as an extra noise even in the absence of bias current. The amplitude of this ferromagnetic resonance-enhanced noise might depend on the probing DC current due to spin transfer torque which affects the quality factor [41].

1.7.6 External Noise

With magnetic sensors, and, in particular very sensitive magnetic sensors, it is essential to take care of the magnetic external noise. Inside a laboratory, there are three types of external perturbations which lead to magnetic noise. First a number of discrete frequencies, including the power supply line (50/60 Hz) and its harmonics but also higher frequencies typically up to MHz coming from power supplies, low consumption lights, etc. All these lines correspond to real AC magnetic fields and hence are proportional to the bias current in the magnetoresistive sensor. The second source is a $1/f$ magnetic noise which exists everywhere. In a laboratory, this noise has an intensity of about 100 nT at 1 Hz and decreases slightly faster than $1/f$. In a good magnetic shielded room with passive and active shielding, the noise level at 1 Hz can be of the order of 100 fT.

The third type of noise is less intuitive. It is created by the vibration of the magnetoresistive sensor in an existing DC field. This noise can only be detected with very sensitive sensors, typically mounted with flux concentrators. The vibration induces a flux, the variation in a flux concentrator, and hence an additional signal. This noise can be easily recognized because it appears usually as bursts at fixed very low frequencies and their amplitude varies strongly when artificial vibrations are created.

1.7.7 Electronics and Noise Measurements

Noise is often difficult to measure quantitatively and there are several approaches to reliably estimate this quantity. Combining field and temperature variation can even complicate the measurement. In this part, we focus on a "standard" setup with some alternatives and with an emphasis on common errors.

1.7.7.1 Electronics Design
The electronics for a noise measurement or for interfacing magnetic sensors in an application can be separated roughly in five parts:

1) The sensor system which may contain one or several different sensors
2) A biasing source (voltage or current)
3) A front end electronics which contains a first preamplifier stage
4) An amplification, filtering, and shaping stage
5) An acquisition system

For noise measurements, the last two parts may be replaced by a spectrum analyzer. Figure 1.6 gives an overview of the electronics for typical noise measurements.

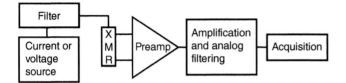

Figure 1.6 Schematics of electronics for typical noise measurements. *Source:* adapted from [32].

1.7.7.2 Connections Noise

In noise investigation as for the development of very sensitive magnetic sensors, it is necessary to take care of the connection quality. Often connections are made with wire bounding in Al or Au with an effective small surface contact. Even if the total resistance contact is small, a low frequency noise can appear. In addition, connector noise along a measurement chain can also be a source of noise which might not be negligible in the total noise if the sensor's noise is low.

1.7.7.3 Correlation for Preamplification Noise Suppression

When the signal of interest is small and dominated by the preamplification noise, it is difficult to measure it. A possible approach consists of using two pre-amplifiers which are measuring the same signal. If a cross-correlation as given in Eq. (1.18) is performed, we obtain the spectral density of the signal of interest plus three terms which are not correlated: the cross-correlation of the signal and each amplifier noise and the correlation between the noise of the two pream-plifiers. By averaging, these three terms are reduced as the square root of the averaging number, but the spectral density is not reduced. Hence, with 100 averages the preamplification noise is reduced by 10. This can be applied for noise studies but not for applications.

References

1 Boll, R. and Overshott, K.J. (2008). *Sensors, Magnetic Sensors*. Wiley.
2 Meyer, H.W. (1971). *A History of Electricity and Magnetism*. Cambridgd, MA: The MIT Press.
3 Schubert, J. (1987). *Dictionary of Effects and Phenomena in Physics: Descriptions, Applications, Tables*. Wiley-VCH.
4 Acuña, M. (1974). Fluxgate magnetometers for outer planets exploration. *IEEE Transactions on Magnetics* 10 (3): 519–523.
5 Heidari, H. (2018). Electronic skins with a global attraction. *Nature Electronics* 1 (11): 578.

6 Williamson, S.J. and Kaufman, L. (1981). Biomagnetism. *Journal of Magnetism and Magnetic Materials* 22 (2): 129–201.

7 Parkinson, D.H. and Mulhall, B.E. (2013). *The Generation of High Magnetic Fields*. Springer.

8 Zijlstra, H. (1967). *Experimental Methods in Magnetism*. North-Holland Pub. Co.

9 Montgomery, D.B. (1969). *Solenoid Magnet Design*. Wiley.

10 Chen, C.-W. (2013). *Magnetism and Metallurgy of Soft Magnetic Materials*. Courier Corporation.

11 Heidari, H., Wacker, N., Roy, S., and Dahiya, R. (2015). Towards bendable CMOS magnetic sensors. In: *2015 11th Conference on PhD Research in Microelectronics and Electronics (PRIME)*, 314–317. IEEE.

12 Nabaei, V., Chandrawati, R., and Heidari, H. (2018). Magnetic biosensors: modelling and simulation. *Biosensors and Bioelectronics* 103: 69–86.

13 Heidari, H., Bonizzoni, E., Gatti, U., and Maloberti, F. (2014). A 0.18-μm CMOS current-mode hall magnetic sensor with very low bias current and high sensitive front-end. In: *SENSORS, 2014 IEEE*, 1467–1470. IEEE.

14 Heidari, H., Bonizzoni, E., Gatti, U. et al. (2016). CMOS vertical Hall magnetic sensors on flexible substrate. *IEEE Sensors Journal* 16 (24): 8736–8743.

15 Bentley, J.P. (1988). *Principles of Measurement Systems*. Pearson Education India.

16 Seippel, R.G. (1983). *Transducers, Sensors & Detectors*. Reston Pub. Co.

17 Korvink, J. and Paul, O. (2010). *MEMS: A Practical Guide of Design, Analysis, and Applications*. Springer Science & Business Media.

18 Middelhoek, S. and Audet, S.A. (1989). *Silicon Sensors*. Academic Press.

19 Sze, S.M. (1994). *Semiconductor Sensors*. New York: Wiley.

20 Roumenin, C.S. (1994). *Solid State Magnetic Sensors*. North-Hollande.

21 Bozorth, R.M. (1993). *Ferromagnetism.* by Richard M. Bozorth, pp. 992. ISBN 0-7803-1032-2. Wiley-VCH, August 1993. 1993 Aug:992.

22 Wohlfarth, E.P. and Buschow, K.H.J. (1980). *Ferromagnetic Materials: A Handbook on the Properties of Magnetically Ordered Substances*. Elsevier.

23 Tebble, R.S., Craik, D.J. (1969). *Magnetic Materials*. London: Wiley-Interscience; 1969 Nov 22.

24 Heck, C. (2013). *Magnetic Materials and Their Applications*. Elsevier.

25 Boll, R. (1979). *Soft Magnetic Materials: Fundamentals, Alloys, Properties, Products, Applications: The Vacuumschmelze Handbook*. Siemens.

26 Brignell, J.E., White, N.M., and Cranny, A.W.J. (1988). Sensor applications of thick-film technology. *IEE Proceedings I-Solid-State and Electron Devices* 135 (4): 77–84.

27 R. S. Popovic (1996). Hall effect devices (Adam Hilger, Bristol, 1991). *Google Scholar*, pp. 165–170

28 Roumenin, C. (1994). Functional magnetic-field microsensors: an overview (S & M 0144). *Sensors and Materials* 5: 285–285.

29 Roumenin, C.S. (1995). Magnetic sensors continue to advance towards perfection. *Sensors and Actuators A: Physical* 46 (1–3): 273–279.

30 Mohri, K., Uchiyama, T., and Panina, L.V. (1997). Recent advances of micro magnetic sensors and sensing application. *Sensors and Actuators A: Physical* 59 (1–3): 1–8.

31 Norton, H.N. (1990). *Sensor and Transducer Selection Guide*. Elsevier Advanced Technology.

32 Reig, C., Cardoso, S., and Mukhopadhyay, S.C. (2013). Giant magnetoresistance (GMR) sensors. *Ssmi6* 1: 157–180.

33 Johnson, J.B. (1927). Thermal agitation of electricity in conductors. *Nature* 119 (2984): 50.

34 Nyquist, H. (1928). Thermal agitation of electric charge in conductors. *Physical Review* 32 (1): 110.

35 Steinbach, A.H., Martinis, J.M., and Devoret, M.H. (1996). Observation of hot-electron shot noise in a metallic resistor. *Physical Review Letters* 76 (20): 3806.

36 Blanter, Y.M. and Büttiker, M. (2000). Shot noise in mesoscopic conductors. *Physics Reports* 336 (1–2): 1–166.

37 Guerrero, R., Aliev, F.G., Tserkovnyak, Y. et al. (2006). Shot noise in magnetic tunnel junctions: evidence for sequential tunneling. *Physical Review Letters* 97 (26): 266602.

38 Garzon, S., Chen, Y., and Webb, R.A. (2007). Enhanced spin-dependent shot noise in magnetic tunnel barriers. *Physica E: Low-Dimensional Systems and Nanostructures* 40 (1): 133–140.

39 Voss, R.F. and Clarke, J. (1976). Flicker ($1/f$) noise: equilibrium temperature and resistance fluctuations. *Physical Review B* 13 (2): 556.

40 Hooge, F.N. (1976). 1/f noise. *Physica B+C* 83 (1): 14–23.

41 Foros, J., Brataas, A., Bauer, G.E.W., and Tserkovnyak, Y. (2009). Noise and dissipation in magnetoelectronic nanostructures. *Physical Review B* 79 (21): 214407.

2

Magnetic Sensors Based on Hall Effect

2.1 Overview

The Hall effect was discovered by Edwin Hall in 1879. Hall found that when a magnet was placed with its field flowing perpendicular to the face of a thin rectangle of gold through which current was flowing, a difference in potential appeared at the opposite edges. He found that this voltage was proportional to the current flowing through the conductor, and the flux density or magnetic induction perpendicular to the conductor. When a current-carrying conductor is placed in a magnetic field, a voltage will be generated, perpendicular to both the current and the field. This principle is known as the *Hall effect*. Figure 2.1 illustrates the basic principle of the Hall effect. It shows a thin sheet of semiconducting material (Hall plate) through which current flows. The output connections are perpendicular to the direction of the current. When no magnetic field is present, the current distribution is uniform and no potential difference is seen across the output. When a perpendicular magnetic field is present, a Lorentz force is exerted on the current. This force disturbs the current distribution, resulting in a potential difference (voltage) across the output. This voltage is the Hall voltage (V_{Hall}).

The magnitude of V_{Hall} is directly proportional to the applied field B and the voltage and current biasing is given by

$$V_{Hall} = G\frac{w}{l}\mu_H V_{bias} B = S_{V_V} V_{bias}B \tag{2.1}$$

$$V_{Hall} = G\frac{r_H}{n.e.t}I_{bias} B = S_{V_I}I_{bias}B \tag{2.2}$$

respectively. Here, G is the geometrical correction factor, w and l stand for the width and length of the plate, μ_H is the Hall mobility of majority carriers, V_{bias} is the total bias voltage, r_H is the Hall scattering factor, n is the carrier concentration, e is the electron charge, t is the thickness of the n-well implantation, I_{bias} is the total bias current, and B is an external magnetic field.

Magnetic Sensors for Biomedical Applications, First Edition. Hadi Heidari and Vahid Nabaei.
© 2020 by The Institute of Electrical and Electronics Engineers, Inc.
Published 2020 by John Wiley & Sons, Inc.

(a) (b)

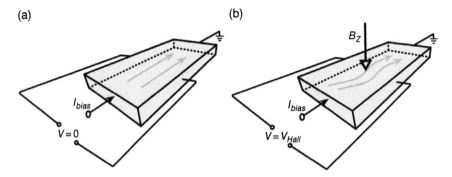

Figure 2.1 A rectangle Hall plate in presence of (a) no magnetic field and (b) a perpendicular magnetic field (B_z).

Moreover, S_{V_V} represents the voltage-related voltage-mode sensitivity and S_{V_I} is the current-related voltage-mode sensitivity, which will be explained in detail in Section 2.5.1. The voltage-related voltage-mode sensitivity S_{V_V} depends on the geometry and the Hall mobility, which is strongly temperature dependent. The current-related voltage-mode sensitivity S_{V_I} is inversely proportional to the carrier concentration n. It is stable in temperature for the plate doping density between 10^{15} and 10^{17} cm^{-3} and in the operating temperature range of many applications nowadays [1]. The current-mode Hall sensor principle is described in the following section. This chapter provides a detailed description of horizontal and vertical Hall plates. The current-mode technique and its comparison with the voltage-mode technique are discussed. Next, the state-of-the-art in Hall magnetic sensors is given. This is followed by an overview of the characteristics of the Hall magnetic sensors in CMOS technology. Applications of the Hall magnetic sensors are presented at the end of the chapter.

2.2 Devices Based on Hall Effect

2.2.1 Geometry

The term Hall Effect devices is used to describe all solid-state electron devices whose principle of operation is based on the Hall effect. Hall devices, which are like the one that Hall had used to discover this effect, today are called *Hall Plates*. Figure 2.2 shows some typical shapes of Hall plates. The bridge shape (c) is a good approximation of a long Hall device and allows relatively large contacts.

The parasitic effects from the contact resistance and the heating can therefore be minimized. The cross-shaped (a) and the Van der Paw shape (b) are of

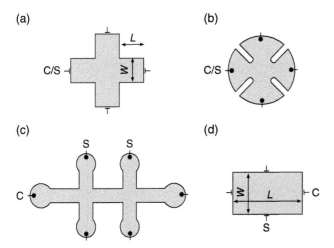

Figure 2.2 Various shapes of Hall plates: (a) Cross-shaped, (b) Van der Paw shape, (c) Bridge shape, and (d) Rectangle shape. C are the current contacts, S the sense contacts, and C/S indicates that the two types of contacts are interchangeable. *Source:* adapted from [1].

interest. They offer the advantages of a high geometrical correction factor ($G1$), a simple and compact geometry, and an invariant shape for rotation through $\pi/2$. The current and sense contacts could be switched without changing the global symmetry. This fourfold symmetry is favorable for the application of off-set voltage compensation techniques like the current spinning technique. This geometry also allows a better definition of the active part center.

2.2.2 Material

The choice of the proper material for the Hall plate is of crucial importance. From Eq. (2.1), one can conclude that the Hall effect will be favored in materials with high mobility and low conductivity. Metals show low mobility and high conductivity and are therefore not a good choice. The sensors are usually prepared from n-type semiconductors where the dominant charge carriers are electrons having much higher mobility than holes. Suitable candidates are intrinsic elements like Si and III–V compounds like InSb, InAs, and GaAs. The III–V compounds combine high carrier mobility and reasonable value of conductivity. Silicon has moderate electron mobility but is compatible with the integrated circuit technology. This makes this material very attractive for the realization of Hall sensor chips.

Table 2.1 gives the values of the energy gap E_g and the mobility at room temperature of various semiconductors used for Hall plates. R_H is also calculated for a given doping density [3].

Table 2.1 Gap and mobility of semiconductors at 300 K used to build a Hall plate [2].

Material	E_g (eV)	μ_n (cm^{-2}V^{-1}·s^{-1})	n (cm^{-3})	R_H (cm^3 C^{-1})^a
Si	1.12	1 500	2.5×10^{15}	2.5×10^3
InSb	0.17	80 000	9×10^{16}	70
InAs	0.36	33 000	5×10^{16}	125
GaAs	1.42	85 000	1.45×10^{15}	2.1×10^3

^{a} R_H is calculated for a given level of doping

2.3 Horizontal Versus Vertical CMOS Hall Devices

The conventional Hall plates are parallel to the chip surface; so, considering a chip as an ocean in which a Hall plate floats, such Hall plates are called *horizontal*. In other words, the horizontal Hall effect device, has a horizontal plate toward the semiconductor substrate and measures the magnetic field, Z, perpendicular to the sensor plate surface, as shown in Figure 2.3a. The active region of the device is compatible with the layers readily available in CMOS technologies. The surface of integrated horizontal Hall devices is in the order of tens of μm^2. Their current-related sensitivities (S_{V_I}) range from 150 to 400 VA^{-1}T^{-1}, while typical voltage-related sensitivities (S_{V_V})^1 are 5–7%T^{-1} [4]. Via special implantation techniques sensitivity values up to S_{V_I} = 2400 VA^{-1}T^{-1} [5] can

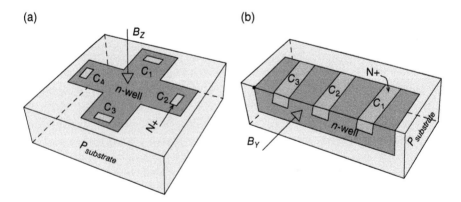

Figure 2.3 (a) Horizontal Hall effect device and (b) three contacts vertical Hall device cross-sections in CMOS technology.

---

1 V/(VT) = T^{-1}

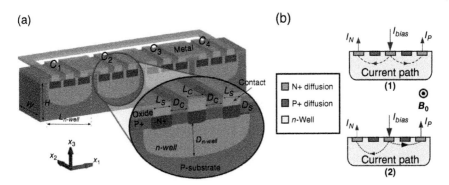

Figure 2.4 Cross-section of a three-contact VHS sensor (a) 3D structure integrated in COMS technology and (b) its current path: without and with lateral magnetic field B_0.
Source: (a) adapted from [9]; (b) adapted from [16].

be obtained. They can be easily co-integrated with signal conditioning electronics in many CMOS Hall effect sensors.

However, a horizontal Hall device, as any Hall device, suffers from offset voltage and low frequency noise. The current spinning method, as a means for offset and low frequency noise reduction, is highly efficient in symmetrical devices. The CMOS Hall sensors based on horizontal Hall devices with current spinning technique feature high resolution, a residual offset as low as 2 µT was reported in [6, 7]. Recently, there has been growing interest in vertical Hall sensors (VHS) [8, 9]. They are sensitive to the in-plane component of the magnetic field, B_Y, and detect the magnetic field in the plane of the sensor plate, as shown in Figure 2.3b, as was first reported in [10]. Several geometries have been studied in [7, 11–13]. However, the sensitivity of vertical Hall devices is low. Values published for voltage-related sensitivity in the literature are 4%T^{-1}, 1.2%T^{-1}, and 1.16%T^{-1}, as reported in [13–15], respectively. Typical current-related sensitivities (S_{V_I}) range from 100 to 400 VA^{-1}T^{-1} is reported in [14]. Various geometries of CMOS vertical Hall devices have also been studied, such as three-contact VHS sensor, and its current path, without and with lateral magnetic field B_0, as shown in Figure 2.4.

2.4 Current-Mode Versus Voltage-Mode Technique

Typically, Hall plates are used in voltage-mode. This means that the magnetic field to be measured is converted into an output voltage. The idea behind a current-mode Hall sensor is to have current and not voltage as an output signal [17–20]. The physical structure of the proposed current-mode Hall sensor is the same for existing devices, with the same possibility of compensating the offset caused by mismatch (current spinning method). The difference is in the way the signals are driven and extracted [21].

Consider the device of Figure 2.5a; the sensor biases current I_{bias} flows from one arm to the other in front (the symmetrical structure is for current spinning). A magnetic field B_z gives rise to the Hall voltage across the two orthogonal arms. Here, it is commonly accepted that the Hall voltage can be calculated from Eq. (2.2). The connection realized in the scheme of Figure 2.5b injects the current laterally in two consecutive arms and a magnetic field causes an unbalance of the two output currents. A difference of these output currents could be represented by an equivalent current source of a Hall current, I_{Hall}. The current-mode Hall sensor principle has been already described in [1], where it has been found that the Hall current is

$$I_{Hall} = \mu_H \frac{w}{l} B I_{bias} \tag{2.3}$$

Here, μ_H is the Hall mobility of majority carriers, I_{bias} is the total bias current, B is a normal magnetic field, and w/l is the width-to-length ratio of the plate. This mode of operation has been extensively studied to estimate the benefit of the current-mode approach. The output currents (I_{H^+}, I_{H^-}) are calculated as

$$I_{H^+} = \frac{I_{bias}}{2} + \frac{I_{Hall}}{2} \tag{2.4}$$

$$I_{H^-} = \frac{I_{bias}}{2} - \frac{I_{Hall}}{2} \tag{2.5}$$

The Hall current (I_{Hall}) is also proportional to the external magnetic field (B_z), biasing current of the Hall plate (I_{bias}) and magnetic resistance coefficient (β). This current, for the cross-shaped Hall plate, can be expressed as:

$$I_{Hall} = \mu_H \frac{\beta B_z I_{bias}}{1 - (\beta B_z)^2} \tag{2.6}$$

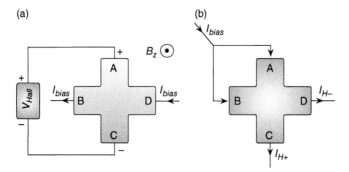

(a)

(b)

Figure 2.5 (a) Hall plate operating in voltage-mode. (b) Hall plate working in current-mode.

where, β is magnetic resistance coefficient in presence of a magnetic field and is calculated as:

$$I_{Hall} = \mu_H \frac{\beta B_z I_{bias}}{1 - (\beta B_z)^2} \tag{2.7}$$

where $R(B_z = 0)$ and $R(B_z)$ define the Hall plate resistance in the absence and presence of an external magnetic field, respectively.

The higher resolution in the low magnetic fields is one of the benefits of the current-mode Hall sensors using high mobility plate. Another benefit is that by using current-mode approach, the number of terminals can be reduced. Therefore, to integrate, the ultimate miniaturization of the system will be easier. Due to the general trend in microelectronics towards miniaturization and thanks to these advantages, the current-mode Hall sensor is bound to become more popular in the future.

2.5 Magnetic Sensors Characteristics

In all applications of Hall magnetic sensors, such as positioning, current sensing, medical and proximity switching, the most important characteristics are magnetic sensitivity, offset-equivalent magnetic field, $1/f$ noise-equivalent magnetic field, signal-to-noise ratio (SNR), and linearity.

For an *AC* magnetic signal above the $1/f$ noise region, the SNRs of equivalent Hall plates in the voltage-mode and current-mode Hall sensor are similar [1]. For a DC and low frequency magnetic field, a voltage-mode Hall sensor in which the current spinning technique is applied can reduce the offset and $1/f$ noise. For the first time an equivalent technique applicable in the current-mode Hall sensor has been presented in [17] and will be thoroughly explained in the following section.

2.5.1 Sensitivity

The magnetic sensitivity is influenced by the biasing conditions. In other words, the magnetic sensitivity of a voltage-biased Hall sensor is proportional to the product (Hall mobility) × (bias voltage), whereas in current-biased Hall sensor it is proportional to the product (Hall coefficient/thickness) × (bias voltage). Thus, current-biased Hall plates are not very temperature-dependent. Indeed, the proposed current-mode Hall sensor is a current-biased circuit and has much smaller temperature dependence compared to voltage-biased Hall sensor [1]. A detailed description of a voltage-mode Hall sensor can be found in [1, 22]. Magnetic Hall sensors are biased in two alternative *voltage-mode* and *current-mode* schemes. In voltage-mode, a sensor is biased by a voltage (V_{bias})

or a current (I_{bias}) and in the presence of a magnetic field, it generates a voltage (V_{Hall}) at the output. This V_{Hall}, as already mentioned in Eq. (2.2), for the bias current can be defined as

$$V_{Hall} = G\frac{r_H}{n.e.t} I_{bias} B \tag{2.8}$$

where n is the doping, r_H the Hall factor, G the geometrical correction factor of Hall voltage, e the elementary carrier charge, B external magnetic field, and t is thickness of the sensor plate. By using $V = Rin \cdot I$, the Hall voltage can be expressed in terms of the bias voltage [1]:

$$V_{Hall} = G\frac{w}{l} \mu_H V_{bias} B \tag{2.9}$$

where, μ_H and w/l denote the Hall mobility of majority carriers and width-to-length ratio of the sensor plate. In case of current-mode condition, the sensor is biased by current and when exposed to an external magnetic field, it produces an output current. This output current, as already discussed, is called the *Hall current* that can be defined from Eq. (2.3).

The absolute sensor sensitivity efficiency for the translation of the magnetic field into the Hall voltage or Hall current depends on the biasing conditions. It is described for voltage-mode by the sensitivity S_{Av} with

$$S_{Av} = \frac{V_{Hall}}{B} [V/T] \tag{2.10}$$

where as absolute sensitivity for current-mode is given by the sensitivity S_{A_I} with

$$S_{A_I} = \frac{I_{Hall}}{B} [A/T] \tag{2.11}$$

The relative sensitivity is defined as the ratio of the absolute sensitivity and a bias quantity (voltage or current). Table 2.2 shows the sensitivity equations for

Table 2.2 Sensitivity of horizontal Hall magnetic sensors in voltage-mode and current-mode conditions.

Mode	Sensitivity	Unit	
Current-biased and voltage-mode	$S_{V_I} = \left\|\dfrac{V_{Hall}}{I_{bias} \times B}\right\|$	$\left[\dfrac{V}{T}\right]$	
Voltage-biased and voltage-mode	$S_{V_V} = \left\|\dfrac{V_{Hall}}{V_{bias} \times B}\right\|$	$\left[\dfrac{V}{VT}\right] = T^{-1}$	
Current-mode	$S_I = \left\|\dfrac{I_{Hall}}{I_{bias} \times B}\right\|$	$\left[\dfrac{A}{AT}\right] = T^{-1}$	

Hall magnetic sensors. Two equations are used to evaluate the performance of the voltage-mode Hall sensor. The first is the current-biased voltage-mode Hall sensitivity, which is proportional to the sensor bias current, I_{bias}, the Hall voltage, V_{Hall}, and the applied perpendicular external magnetic field, B. The units are $V \cdot A^{-1} \cdot T^{-1}$ (volt/(ampere·Tesla)). The second equation describes the efficiency of a Hall device by its voltage-biased voltage-mode Hall sensitivity S_{V_V} in terms of *per tesla* $(V/(V \cdot T)) = T^{-1}$. In this equation, V_{bias} denotes the supplied bias voltage. The voltage-biased voltage-mode sensitivity S_{V_V} can be defined by

$$S_{V_V} = \frac{S_{V_I}}{R_{in}} \tag{2.12}$$

The current-mode magnetic sensitivity is used to evaluate the performance of current-mode Hall sensors, which is defined as the third equation in Table 2.2. Here I_{bias} is the sensor bias current, I_{Hall}, the Hall current. In this way, the units are also defined *per tesla* $(A/(A \cdot T)) = T^{-1}$ and are the same as in voltage-biased voltage-mode units.

Current-mode Hall sensors show better sensitivity compared to voltage-mode [6]. To show the efficiency of current-mode approach, a Wheatstone bridge model of cross-shape Hall sensor has been used. The applied magnetic field alters the current flow in the body sensor and this gives rise to a variation of the sheet to body resistance in the device. At the macro level, we can model the sensor as a bridge of resistances as shown in Figure 2.6. Four resistors model the

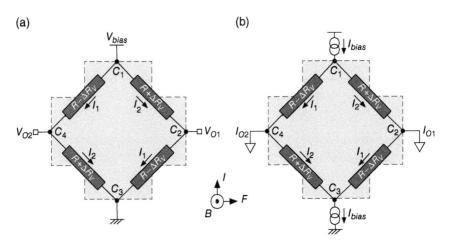

Figure 2.6 Wheatstone bridge model of cross-shape Hall sensor in (a) voltage-mode and (b) current-mode configuration.

DC electrical relationship between the input and output nodes. The applied magnetic field changes the value of resistors by ΔR_V.

Since the voltage-mode and the current-mode of operation define different boundary conditions, the magnetic field would affect differently the current flow in the device and, therefore, the bridge resistors' variation. The study requires 2D simulations. Here, to obtain first approximation results, we suppose the same resistor change in both modes of operation.

For the voltage-mode case (Figure 2.6a) it results

$$V_{O1} = \frac{V_{bias}}{2R}(R - \Delta R_V) \tag{2.13}$$

$$V_{O2} = \frac{V_{bias}}{2R}(R + \Delta R_V) \tag{2.14}$$

The differential Hall voltage is given by

$$V_{Hall} = V_{bias}\frac{\Delta R_V}{R} \tag{2.15}$$

For the current-mode Hall sensors (Figure 2.6b) a bias current is injected into a terminal (C_1) of the Hall plate, and the same bias current is drawn from opposite terminal (C_3). The output terminals (C_2 and C_4) are at the same ground voltage. The differential output currents are given by

$$I_1 = \frac{V_{C1}}{R - \Delta R_1} \tag{2.16}$$

$$I_2 = \frac{V_{C1}}{R + \Delta R_1} \tag{2.17}$$

that yield the differential Hall current as

$$I_{Hall} = I_{bias} \cdot 2 \cdot \frac{\Delta R_I}{R} \tag{2.18}$$

Therefore, the sensitivity in the current-mode with equal change of the resistors in the model is twice larger than the voltage-mode. There is another side benefit in having current as output variable. Its integration over a capacitance for a defined period of time determines an output voltage with a gain factor proportional to the integration time. This is a simple way to amplify the signal by using a single fully differential operational amplifier.

A possible mismatch of the voltages of the output nodes causes an offset in the output signal. This limit can be compensated for by a suitable control in the electronic interface.

2.5.2 Offset

The offset of a Hall device is the output voltage when no magnetic field is present $V_{OS} = V_{out}$ (B = 0 T). According to sensitivity (Table 2.2), the offset equivalent magnetic field, B_{OS}, is defined by

$$B_{OS} = \frac{V_{OS}}{Absolute\ sensitivity}\ [T] \tag{2.19}$$

For the direct measurement of this value, the sensor should be put into a shielded region to exclude all present magnetic fields. The main causes of offset are related to a structural asymmetry of the active part (errors in geometry, nonuniform doping density, contact resistance, etc.). Furthermore, the piezoresistance effect and alignment errors of the sense contacts influence offset value. The typical value of offset equivalent magnetic field for a Hall device fabricated in a silicon integrated technology is 5–50 mT [7].

To reduce the offset of the output signal, there are several techniques such as orthogonal coupling of Hall devices and current spinning. The orthogonal coupling of Hall devices is based on pairing of an even number of Hall devices and biasing them orthogonally [23], as shown in Figure 2.7a for two cross-shaped Hall plates [4]. The offset voltage is dependent on geometry of a Hall device, whereas the Hall voltage and subsequently sensitivity are independent

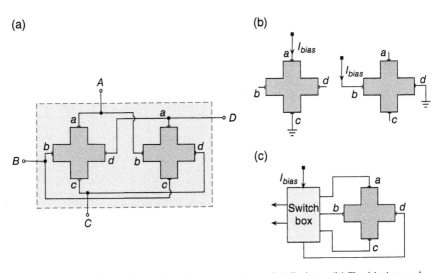

Figure 2.7 (a) Orthogonal coupling of two cross-shaped Hall plates. (b) The biasing and sensing are altered using the symmetry of the Hall plate. (c) The current spinning technique: the biasing and sensing contacts are switched periodically to modulate either the Hall voltage or the offset. *Source:* adapted from [8].

functions. As a result, the outputs of the devices can be connected, so that the respective Hall voltages are averaged while offset voltages are canceled out.

In Section 2.5.2.1, another method of offset voltage cancelation, named *current spinning* technique, is described. It is also known as the connection-commutation, switched Hall plate, or dynamic method. The orthogonal coupling and current spinning technique can be combined in an array of Hall devices, as presented in [24]. The remaining offset after the offset reduction techniques are applied is referred to as residual offset.

2.5.2.1 Current Spinning Technique

The current spinning technique allows to strongly reduce the offset of Hall sensors [10]. The use of the symmetric Hall plate enables to apply the current spinning technique. As shown in Figure 2.7b, the current spinning interchanges periodically the output and supply terminals of the Hall plate so that the input bias injecting point is rotated in each state, whereas the offset appears at the output terminals. The plate is clocked with several phases and the output signals are summed. Normally, the spinning switches for the periodic change of biasing and sensing contacts are known as the switch box, as shown in Figure 2.7c. Using this technique, the Hall plate offset can be reduced to about 50–100 µT.

Depending on the number of different biasing phases there are a two-phase spinning system using two different biasing states of the sensor, a four-phase spinning using four phases, or even an eight-phase spinning with eight biasing modes. Figure 2.8a shows an implementation of current spinning in a four-contact Hall sensor with four phases, reported in [25]. Figure 2.8b illustrates an octagonal Hall plate that required the eight-phase current spinning, reprinted from [26]. For CMOS integrated Hall plates, offset values of several hundreds of µT are generally reported for a two-phase current spinning circuit. For four-phase and eight-phase current spinnings, offset fields of 25 and 5 µT are reported in [25, 26], respectively. To further reduce the offset of the Hall plate, four spinning-current Hall plates were orthogonally coupled to form a quad Hall plate structure in [6], as shown in Figure 2.9. Orthogonally coupled Hall plates are parallel-coupled Hall plates in both current supporting contacts and Hall voltage contacts; however, the current direction in the Hall plates differs by 90°. Such a structure instantaneously compensates for offset due to mechanical stress, and combined with the spinning-current method, reduces the offset to 3.65 µT.

The current spinning technique can also compensate the $1/f$ noise of the sensor. Usually the switching frequency should be bigger than the noise corner frequency to obtain a good compensation of the $1/f$ noise. The possibility to apply the current spinning to VHS has been reported in [27, 28]. Figure 2.10 shows the four phases for a five-contact vertical Hall sensor.

(a)

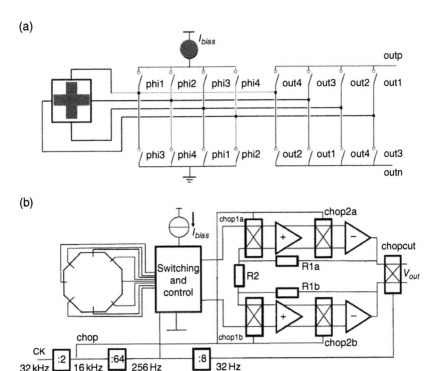

(b)

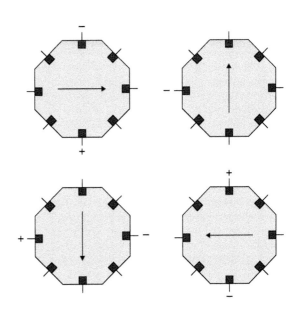

Figure 2.8 (a) Applying the current spinning technique to four-contact and (b) eight-contact horizontal Hall sensors. *Source:* adapted from [25, 26].

Figure 2.9 Four phases of a quad current spinning Hall plate to reduce the offset. *Source:* adapted from [6].

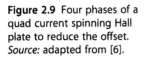

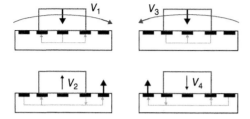

Figure 2.10 Spinning-current technique applying to five-contact vertical Hall sensor. *Source:* adapted from [27].

2.5.3 Noise

The output voltage of a Hall sensor shows, in general, random fluctuations called noise [29]. This means that a noise voltage V_N is superposed with the Hall voltage and sensor offset. The sensor output $Vout$ therefore becomes:

$$V_{out} = V_{Hall} + V_{off} + V_N \qquad (2.20)$$

The noise in a Hall sensor is due to thermal noise, generation-recombination noise, and $1/f$ noise [1]. The main sources of noise are internal, for example, resistance of the sensor is the source of *thermal Johnson noise* V_{nT} while semiconductor junction is the source of shot noise I_{ns}:

$$V_{nT} = \sqrt{(4kTR\Delta f)} \qquad (2.21)$$

$$I_{ns} = \sqrt{(2qI\Delta f)} \qquad (2.22)$$

where k is the Boltzmann constant and q is the electron charge [28].

Due to the fact that the noise depends on the frequency range Δf, usually the spectral density $S(f)$ of the noise is defined as

$$S(f) = \frac{V_n^2}{\Delta f} = \left(\frac{V_n}{\sqrt{\Delta f}}\right)^2 \qquad (2.23)$$

Therefore, a *unit* of noise can be determined as V^2/Hz, or a noise equivalent magnetic field as $nT/\sqrt{Hz}$. A convenient way to describe the noise properties of a magnetic sensor is in terms of resolution, which is also called detection limit. A high resolution can be achieved in a large Hall device made of high-mobility material and low $1/f$ noise parameter when it operates at high power level. These noise can be minimized with good laboratory practice and experiment design or through suitable compensation techniques.

2.5.4 Nonlinearity

Magnetic Hall sensors show an almost linear response for magnetic fields up to 1 T. The nonlinearity describes the real measurement result deviation from the best ideal linear fit. It is used to evaluate the accuracy of the sensor. A good review of nonlinearity problems on Hall sensors is given in [30, 31].

The nonlinearity in the Hall sensors is due to current-related sensitivity, material nonlinearity, geometrical nonlinearity, and nonlinearity caused by the junction field effect. The current-related nonlinearity will appear in a magnetic sensor if its sensitivity depends on the magnetic field. The junction field effect nonlinearity depends on the device structure, biasing conditions, and the magnetic field [11].

The nonlinearity error (NLE) in the voltage-mode Hall sensor is also defined as

$$NLE_V = \left| \frac{V_{Hall} - V_{Hall}(0)}{V_{Hall}(0)} \right| \times 100\% \tag{2.24}$$

where V_{Hall} is the measured value of the Hall voltage and $V_{Hall}(0)$ is the calculated value based on the slope of the straight line obtained by the best fit to the output characteristic.

In the same way, the NLE in the current-mode Hall sensor can be specified as

$$NLE_I = \left| \frac{I_{Hall} - I_{Hall}(0)}{I_{Hall}(0)} \right| \times 100\% \tag{2.25}$$

where I_{Hall} is the measured value and $I_{Hall}(0)$ is the calculated value of the Hall current.

2.6 State-of-the-art in CMOS Hall Magnetic Sensors

Hall effect sensors have been the workhorse magnetic sensor for decades. Over the last few years, two major achievements were carried out in the development of Hall sensors in CMOS technology: increase of magnetic sensitivity and reduction of offset. Therefore, the future of the Hall devices will depend on the *sensitivity improvement* and *offset reduction*. These two achievements are discussed in detail for vertical and horizontal Hall magnetic sensors in this section.

2.6.1 Sensitivity Improvement

The sensitivity performance of vertical and horizontal Hall sensor has already been explained in Section 2.5.1 [32]. In horizontal Hall sensors, current-biased voltage-mode sensitivities (S_{V_I}) range from 150 to 400 $VA^{-1}T^{-1}$, while typical voltage-biased voltage-mode sensitivities (S_{V_V}) are 2.5–5%T^{-1}.

On the other hand, VHS presented in literature exhibit current-biased voltage-mode sensitivities (S_{V_I}) ranging from 100 to 400 $VA^{-1}T^{-1}$, and typical voltage-biased voltage-mode sensitivities (S_{V_V}) are 1–4%T^{-1}, all them dealing

Table 2.3 Sensitivity comparison of horizontal and vertical Hall magnetic sensors.

References	Sensor type	Sensor plate	Sensitivity (%T)
[33]	Horizontal	Five strips	4.6
[34]	Horizontal	n-Well and MOS	3
[35]	Horizontal	n-Well resistor	2.5
[36]	Horizontal	n-Well	2.9
[37]	Horizontal	Octagonal	4.5
[12]	Vertical	5-Contact	1.52
[15]	Vertical	3C Four-folded	1.72
[13]	Vertical	2D 5-Contact	4.3
[14]	Vertical	5-Contact	4

with voltage as output quantity. Various geometries of CMOS horizontal and vertical Hall devices have been studied and investigated for increasing the sensitivity, such as [33–37] for the horizontal case, and three-contact four-folded [15], four-contact (4C) [38], five-contact (5C) [10], six-contact (6C) [13, 14] for vertical Hall devices. Table 2.3 summarizes the sensitivity performances for state-of-the-art Hall devices.

2.6.2 Offset Reduction

Some of state-of-the-art works that use the current spinning technique to achieve offset reduction have already been discussed in Section 2.5.2. The offset of a Hall effect magnetic sensor system is the sum of the uncanceled offsets of a Hall device and the signal conditioning electronics. The typical offset equivalent magnetic field of a Hall device realized in silicon integrated technology is 5–50 mT. It can be reduced by some techniques, such as current spinning, to lower than 100 μT.

Table 2.4 summarizes and compares most low-offset CMOS Hall sensors reported in literature. In [25], sensor offset has been canceled before amplification by means of a triple ripple reduction loops (RRL) scheme. The offset of Hall plates in [6] is reduced to the 10 μT level by the current spinning method and is also instantaneously reduced by orthogonally coupling four Hall-plates. Again in [39], by the current spinning principle, the signal voltage is separated by the offset voltage through modulation. Therefore, Hall plate offset and amplifier offset are suppressed. Chopping and current spinning techniques have also been used in [40] to reduce the influence of the undesired offset. A current spinning Hall scheme and analog chopping is used in [41] to reduce the offset.

Table 2.4 Offset comparison of horizontal and vertical Hall magnetic sensors.

References	Sensor principle	Offset (μT)
[25]	Orthogonally parallel	25
[6]	Four orthogonally coupled	3.65
[39]	Five-octagon IMC	10
[40]	Two pairs of orthogonally coupled	<200
[41]	Differential Hall probes	10

2.7 Applications of Hall Magnetic Sensors

Sensing the presence or position of an object are two of the most widespread applications in which Hall sensors are used [42–45]. Magnetic field sensors are well suited to this kind of application. In the following, several ways to use Hall magnetic sensors to perform a sensing function will be discussed.

One of the major drivers in this market is the increasing usage of magnetic sensors in automotive and electronic compasses. The Automotive and Electronics industries are the largest end-users of magnetic sensors. Magnetic sensors provide safety as they are used in power trains, air bags, air pressure controls, and fuel systems. The increasing demand for smartphones and tablets also has a positive impact on the market.

In this section, a few examples of Hall magnetic sensors are presented such as biosensors, contactless current sensors, angular position sensor, linear position sensor, contactless joystick sensor, electronic compass, speed and timing sensors, and specific sensors. More information on the most important application can be found in the literature of the providers of Hall magnetic sensors. Practically all manufacturers of Hall sensors (Asahi Kasei Microsystems of Japan, Massachusetts-based Allegro Microsystems, Infineon Technologies of Germany, Micronas of Switzerland, and Belgium-based Melexis) offer a large variety of applications [46–50].

2.7.1 Biosensors

Magnetic Hall sensors are suitable for integration and scaling in CMOS technology, and hence, enable very compact and low-cost systems. Magnetically labeled bioassays are therefore a promising approach for addressing the requirements and growing need of point-of-care diagnosis [36]. Figure 2.11 shows the cross section of a magnetic bead sensor plate which is used for detecting the magnetic labels.

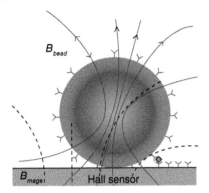

Figure 2.11 Cross-section of a magnetic bead (label) detector using CMOS magnetic Hall sensor. *Source:* adapted from [36].

For measuring magnetic field produced by electro- and permanent magnets, in the range of 10^{-4} µT to a few tens of tesla, Hall sensors are typically used [1]. For example, they are used to test the electromagnets applied in particle accelerators, nuclear magnetic resonance (NMR) imaging systems, and NMR spectroscopy systems.

2.7.2 Contactless Current Sensors

Measuring and sensing electrical current can be done indirectly, by measuring the magnetic field associated with the current flow. The main advantage of magnetic current sensing is that it does not interfere with the circuit in which the current is being sensed [42]. A basic configuration of a contactless current Hall effect sensor has been shown in Figure 2.12. The three Hall effect-based technologies have been used for AC and DC current measurement. First, open-loop Hall effect transducers use a Hall sensing plate placed into the air gap of a magnetic circuit. The second technology is based on the closed-loop current sensing. The difference between the open-loop and the closed-loop current transducers is that the latter has a built-in compensation circuit which improves performance. Finally, Hall effect ETA transducers employ a combination of open-loop and closed-loop technologies. At low frequencies (2–10 KHz) they work as open-loop transducers; thus the Hall plate is providing a signal proportional to the primary current to be measured. At high frequencies, they work as current transformers. These signals are electronically added to form a common output signal [51].

2.7.3 Contactless Angular, Linear, and Joystick Position Sensors

An angular position sensor (also referred to as a rotary sensor) measures the relation by which any position with respect to any other position is established.

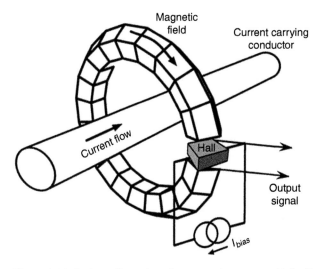

Figure 2.12 Basic configuration of a contactless current Hall effect sensor.

It calculates the orientation of an object with respect to a specified reference position as expressed by the amount of rotation necessary to change from one orientation to the other about a specified axis. The magnetization of the magnet is perpendicular to the axis and parallel with the sensor plane. With a rotation of the axis the flux density vector rotates in the sensor plane and the output signal of the two sensor axes X and Y yields sine and cosine signals with the rotation [52].

For linear position measurement, the two-axis Hall sensor is combined with a magnet, which is now magnetized orthogonally to the sensor plane. The magnet is now shifted parallel to one axis at a certain distance above and sideways of the sensor. This again leads to a sine and cosine reading of the direction of the field angle. The ratio of both values is directly proportional to the magnet position [52].

By using all three-magnetic field sensitive axes, a contactless magnetic joystick can be implemented. The two tilt angles in the XZ and YZ planes can be computed very accurately by using all three-magnetic field components. Here again, the position information is derived from the ratio of magnetic field components, so that temperature and ageing drift of magnet and sensor are canceled out [52].

2.7.4 Electronic Compass

The Earth's magnetic field intensity is about 50 μT and has a component parallel to the Earth's surface that always points towards the Earth's magnetic north. This is the basis for all magnetic compasses. Nowadays, various types of

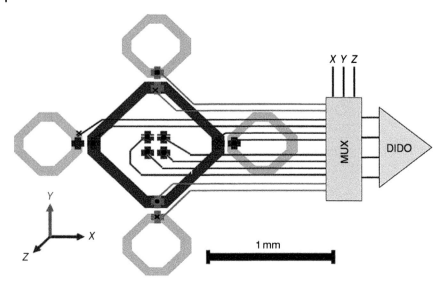

Figure 2.13 Entire IMC structure and Hall plates that combines the advantages of simple post-processing and low-power consumption. *Source:* adapted from [53].

technologies are applied to electronic compasses, and they are based on the magnetoresistive (e.g. anisotropic magnetoresistance [AMR] or giant magneto-resistance [GMR]) effect, the fluxgate effect, or on the magneto-impedance (MI) effect in zero-magnetostrictive amorphous wires. Although these sensors are more sensitive and temperature-stable with higher bandwidth, they either require a lot of power or a very delicate postprocessing of the ferromagnetic core. In [39], a Hall-IMC compass is presented that combines the advantages of simple postprocessing and low-power consumption, as shown in Figure 2.13. For realizing this compass, four Hall plates have been used simultaneously and their outputs are added in an amplifier.

2.7.5   Speed and Timing Sensors

The ability to measure the speed or position of a rotating shaft is necessary for the proper function of many types of machinery. Hall sensors are commonly used to time the speed of wheels and shafts, such as for internal combustion engine ignition timing, tachometers, and antilock braking systems (ABS) [42].

2.7.6   Specific Sensors

Some of the Hall sensors are used as application-specific sensors. For example, Melexis designed a Hall IC to provide control and power-driver functions for a small brushless DC motor. This type of motor is commonly used in the cooling fans found in virtually all personal computers. The Hall sensor is also used in the brushless DC motor to sense the position of the rotor and to switch the transistor in the right sequence.

References

1 Sze, S.M. (2008). *Semiconductor Devices: Physics and Technology*. Wiley.

2 Kvitkovic, J. (1997). Hall generators. *CERN Accelerator School on Measurement and Alignment of Accelerator and Detector Magnets*, Anacapri, Italy (11–17 April 1997; CERN-98-05), pp. 233–249.

3 Sanfilippo, S. (2011). Hall probes: physics and application to magnetometry. CERN-2010-004, pp. 423–462. arXiv:1103.1271.

4 Demierre, M. (2003). Improvements of CMOS Hall microsystems and application for absolute angular position measurements. PhD, EPFL.

5 Randjelovic, Z., Popovic, R., Petr, J. et al. (1999). Low power high sensitive CMOS magnetic field Hall sensor. *Proceedings of the Eurosensors XIII, 15B3*, The Hague, the Netherlands, pp. 601–604.

6 Van Der Meer, J.C., Riedijk, F.R., van Kampen, E. et al. (2005). A fully integrated CMOS Hall sensor with a 3.65/spl mu/T 3/spl sigma/offset for compass applications. In: *ISSCC. 2005 IEEE International Digest of Technical Papers. Solid-State Circuits Conference, 2005*, 246–247. IEEE.

7 Banjevic, M. (2011). High bandwidth CMOS magnetic sensors based on the miniaturized circular vertical Hall device. PhD, EPFL.

8 Heidari, H., Bonizzoni, E., Gatti, U., and Maloberti, F. (2014). Analysis and modeling of four-folded vertical Hall devices in current domain. In: *2014 10th Conference on Ph.D. Research in Microelectronics and Electronics (PRIME)*, 1–4. IEEE.

9 Heidari, H., Bonizzoni, E., Gatti, U. et al. (2016). CMOS vertical Hall magnetic sensors on flexible substrate. *IEEE Sensors Journal* 16 (24): 8736–8743.

10 Popovic, R. (1984). The vertical Hall-effect device. *IEEE Electron Device Letters* 5 (9): 357–358.

11 Kaufmann, T., Vecchi, M., Ruther, P., and Paul, O. (2012). A computationally efficient numerical model of the offset of CMOS-integrated vertical Hall devices. *Sensors and Actuators A: Physical* 178: 1–9.

12 Paul, O., Raz, R., and Kaufmann, T. (2012). Analysis of the offset of semiconductor vertical Hall devices. *Sensors and Actuators A: Physical* 174: 24–32.

13 Sung, G.-M. and Yu, C.-P. (2013). 2-D differential folded vertical Hall device fabricated on a p-type substrate using CMOS technology. *IEEE Sensors Journal* 13 (6): 2253–2262.

14 Schurig, E. (2005). Highly sensitive vertical Hall sensors in CMOS technology. PhD, EPFL.

15 Sander, C., Raz, R., Ruther, P. et al. (2013). Fully symmetric vertical Hall devices in CMOS technology. In: *SENSORS, 2013 IEEE*, 1–4. IEEE.

16 Lei, K.-M., Heidari, H., Mak, P.-I. et al. (2017). A handheld high-sensitivity micro-NMR CMOS platform with B-field stabilization for multi-type biological/chemical assays. *IEEE Journal of Solid-State Circuits* 52 (1): 284–297.

17 Heidari, H., Bonizzoni, E., Gatti, U., and Maloberti, F. (2014). A current-mode CMOS integrated microsystem for current spinning magnetic Hall sensors. In: *2014 IEEE International Symposium on Circuits and Systems (ISCAS)*, 678–681. IEEE.

18 Heidari, H., Bonizzoni, E., Gatti, U., and Maloberti, F. (2014). A 0.18-μm CMOS current-mode Hall magnetic sensor with very low bias current and high sensitive front-end. In: *SENSORS, 2014 IEEE*, 1467–1470. IEEE.

19 Heidari, H., Bonizzoni, E., Gatti, U. et al. (2015). A CMOS current-mode magnetic Hall sensor with integrated front-end. *IEEE Transactions on Circuits and Systems I: Regular Papers* 62 (5): 1270–1278.

20 Heidari, H., Zuo, S., Krasoulis, A., and Nazarpour, K. (2018). CMOS magnetic sensors for wearable magnetomyography. *40th International Conference of the IEEE Engineering in Medicine and Biology Society (EMBC)*, vol. 37, no. 4: 2116–2118.

21 Heidari, H., Wacker, N., Roy, S., and Dahiya, R. (2015). Towards bendable CMOS magnetic sensors. In: *2015 11th Conference on Ph.D. Research in Microelectronics and Electronics (PRIME)*, 314–317. IEEE.

22 Pastre, M., Kayal, M., and Blanchard, H. (2007). A Hall sensor analog front end for current measurement with continuous gain calibration. *IEEE Sensors Journal* 7 (5): 860–867.

23 Maupin, J.T. and Geske, M.L. (1980). The Hall effect in silicon circuits. In: *The Hall Effect and Its Applications* (ed. C.L. Chien and C.R. Westgate), 421–445. Boston, MA: Springer.

24 Hohe, H.-P., Weber, N., and Seitzer, D. (1999). Sensor topology for offset reduction in hall-effect devices. *Proceedings of the Eurosensors XIII*, The Hague, the Netherlands, pp. 605–608.

25 Jiang, J., Kindt, W.J., and Makinwa, K.A. (2014). A continuous-time ripple reduction technique for spinning-current Hall sensors. *IEEE Journal of Solid-State Circuits* 49 (7): 1525–1534.

26 Bakker, A., Bellekom, A., Middelhoek, S., and Huijsing, J. (1999). Low-offset low-noise 3.5 mW CMOS spinning-current Hall effect sensor with integrated chopper amplifier. *Proceedings of the Eurosensors XIII*, The Hague, the Netherlands, pp. 1045–1048.

27 Madec, M., Osberger, L., and Hebrard, L. (2013). Assessment of the spinning-current efficiency in cancelling the 1/f noise of vertical Hall devices through accurate FEM modeling. In: *SENSORS, 2013 IEEE*, 1–4. IEEE.

28 Tumanski, S. (2016). *Handbook of Magnetic Measurements*. CRC press.

29 Heidari, H., Gatti, U., Bonizzoni, E., and Maloberti, F. (2013). Low-noise low-offset current-mode Hall sensors. In: *Proceedings of the 2013 9th Conference on Ph.D. Research in Microelectronics and Electronics (PRIME)*, 325–328. IEEE.

30 Popović, R. and Hälg, B. (1988). Nonlinearity in Hall devices and its compensation. *Solid-State Electronics* 31 (12): 1681–1688.

31 Caruntu, G. and Dragomirescu, O. (2001). The nonlinearity of magnetic sensors. In: *2001 International Semiconductor Conference: CAS 2001 Proceedings*, vol. 2, 375–378. IEEE.

32 Heidari, H., Gatti, U., and Maloberti, F. (2015). Sensitivity characteristics of horizontal and vertical Hall sensors in the voltage- and current-mode. In: *2015 11th Conference on Ph.D. Research in Microelectronics and Electronics (PRIME)*, 330–333. IEEE.

33 Kammerer, J.-B., Hébrard, L., Frick, V. et al. (2003). Horizontal Hall effect sensor with high maximum absolute sensitivity. *IEEE Sensors Journal* 3 (6): 700–707.

34 Skucha, K., Liu, P., Megens, M. et al. (2011). A compact Hall-effect sensor array for the detection and imaging of single magnetic beads in biomedical assays. In: *16th International Solid-State Sensors, Actuators and Microsystems Conference (TRANSDUCERS)*, 1833–1836. IEEE.

35 Gambini, S., Skucha, K., Liu, P.P. et al. (2013). A 10 kPixel CMOS hall sensor array with baseline suppression and parallel readout for immunoassays. *IEEE Journal of Solid-State Circuits* 48 (1): 302–317.

36 Liu, P.P., Skucha, K., Duan, Y. et al. (2012). Magnetic relaxation detector for microbead labels. *IEEE Journal of Solid-State Circuits* 47 (4): 1056–1064.

37 Ramirez, J.L. and Fruett, F. (2013). Octagonal geometry Hall plate designed for the PiezoHall effect measurement. In: *28th Symposium on Microelectronics Technology and Devices (SBMicro 2013)*, 2013 Sep 2, pp. 1–4. IEEE.

38 Falk, U. (1990). A symmetrical vertical Hall-effect device. *Sensors and Actuators A: Physical* 22 (1-3): 751–753.

39 Schott, C., Racz, R., Manco, A., and Simonne, N. (2007). CMOS single-chip electronic compass with microcontroller. *IEEE Journal of Solid-State Circuits* 42 (12): 2923–2933.

40 Motz, M., Draxelmayr, D., Werth, T., and Forster, B. (2005). A chopped hall sensor with small jitter and programmable "true power-on" function. *IEEE Journal of Solid-State Circuits* 40 (7): 1533–1540.

41 Motz, M., Ausserlechner, U., Bresch, M. et al. (2012). A miniature digital current sensor with differential Hall probes using enhanced chopping techniques and mechanical stress compensation. In: *Sensors, 2012 IEEE*, 1–4. IEEE.

42 Ramsden, E. (2011). *Hall-Effect Sensors: Theory and Application*. Elsevier.

43 Manzin, A. and Nabaei, V. (2014). Modelling of micro-Hall sensors for magnetization imaging. *Journal of Applied Physics* 115 (17): 17E506.

44 Manzin, A., Nabaei, V., and Kazakova, O. (2012). Modelling and optimization of submicron Hall sensors for the detection of superparamagnetic beads. *Journal of Applied Physics* 111 (7): 07E513.

45 Nabaei, V., Rajkumar, R., Manzin, A. et al. (2013). Optimization of Hall bar response to localized magnetic and electric fields. *Journal of Applied Physics* 113 (6): 064504.

46 Asahi Kasei Microdevices (AKM). www.akm.com (accessed 16 August 2019).

47 Allegro Microsystems. www.allegromicro.com (accessed 16 August 2019).

48 Infineon Technologies. www.infineon.com (accessed 16 August 2019).

49 Micronas. www.micronas.com (accessed 16 August 2019).

50 Melexis. www.melexis.com (accessed 16 August 2019).

51 Liakou, F. (2012). Galvanically isolated wide-band current sensors. PhD, University of Thessaly.

52 Popovic, R.S. (2010). *Hall Effect Devices*, 2e. CRC Press.

53 Schott, C., Racz, R., Huber, S. et al. (2008). CMOS single-chip electronic compass with microcontroller. In: *Analog Circuit Design* (ed. H. Casier, M. Steyaert, and A.H.M. Van Roermund), 55–69. Springer.

3

Magnetoresistive Sensors

3.1 Introduction

Magnetoresistive (MR) sensors are linear magnetic field transducers based either on the intrinsic magnetoresistance of the ferromagnetic (FM) material (sensors based on the spontaneous resistance anisotropy in 3D FM alloys, also called anisotropic magnetoresistance [AMR] sensors) or on FM/nonmagnetic heterostructures (giant magnetoresistance [GMR] multilayers, spin valve, and tunneling magnetoresistance [TMR] devices) [1]. MR sensors are widely used in various applications. The most important are their applications as read heads. MR sensors are also used as mechanical transducers and of course as magnetic field sensors, for example, in compasses. The progress in thin film MR sensors is possible due to the large scientific and business interest. It has always been to the benefit of MR sensors that thin film technology and magnetoresistance were the objects of interest of other often better financial supported activities. For fundamental physics, the investigations of phenomena in thin magnetic films were always valuable because it was easier to explain and to understand the magnetism in this specific form of magnetic material (in comparison with the bulk form). Usually MR sensors are divided into AMR and GMR sensors. This classification results from the different mechanisms and features of these effects. The GMR is recently in vogue partially due to the very marketing name. Simultaneously the AMR sensors are treated as slightly old-fashioned. Meanwhile the AMR sensors still exhibit better sensitivity and are much simpler in technology. In 1857, William Thomson (Lord Kelvin) first reported the AMR effect, observing that when iron was subjected to a magnetic field there was a 0.033% increase in its electrical resistance [2]. This very subtle effect is the result of the variation of electron mean free path as a function of the angle of the electron velocity with respect to material magnetization.

In other words, the physical origin of AMR effect, which is characteristic of transition FM materials (and their alloys), lies in the spin-orbit coupling, reflecting the interaction between the spin of the conduction electrons and the crystal

Magnetic Sensors for Biomedical Applications, First Edition. Hadi Heidari and Vahid Nabaei.
© 2020 by The Institute of Electrical and Electronics Engineers, Inc.
Published 2020 by John Wiley & Sons, Inc.

lattice. As the magnetization rotates, the conduction electrons undergo a different amount of scattering when traversing the lattice; macroscopically this effect is seen as a change in the electrical resistance of the material [3]. The sensitivity of MR materials is expressed as the change in resistance divided by the minimum resistance (MR ratio).

3.2 Materials and Principles of AMR, GMR, and TMR

3.2.1 Anisotropic Magnetoresistance

3.2.1.1 Anisotropic Magnetoresistance Effect and Principles

The AMR effect leads to a local dependence of electrical conductivity on the mutual orientation of magnetization and current density vectors. The FM sample shows high resistivity when the magnetization direction is parallel to the current and low resistivity when they are perpendicular [4, 5].

In general, magnetic field can affect the conduction of substances in several ways, giving rise to several so-called galvanomagnetic effects. A comprehensive summary of these effects has been presented in [6]. In ferromagnetics, there is a strong influence of the material's own magnetization, besides the external magnetic field, making the situation more complicated. Due to their specificity, the galvanomagnetic effects in ferromagnetics are sometimes called the extraordinary galvanomagnetic effects; the effect observed by William Thomson, called extraordinary or AMR, being one of them [7].

MR effect may be a consequence of the Lorentz force–similarly as in the Hall effect [8]. The Lorentz force F is acting on particle with charge q moving with velocity v in a magnetic field B

$$F = q(v \times B) \tag{3.1}$$

As a result, the Lorentz force deflects the path of the moving free charge carrier (Figure 3.1a). It increases the path length causing a change in the material resistivity. Figure 3.1b and c present the current paths and equipotential lines in a rectangular plate of the conducting material.

The resistance $R(B)$ is proportional to the square of the magnetic field induction B perpendicular to the plate:

$$R(B) = R_0 \frac{\rho_B}{\rho_0} \left(1 + \mu^2 C B^2\right) \tag{3.2}$$

where R_0 is the resistance of the material in the magnetic field $B = 0$, ρ_B/ρ_o is the magnetoresistivity coefficient, μ is the carrier mobility, and C is the coefficient depending on the geometry of the sample. The carrier mobility is especially large in semiconductors InSb and InAs and therefore these materials are most frequently used to prepare the magnetoresistors ($\mu = 7.7 \times 10^4$ cm^2/Vs for InSb

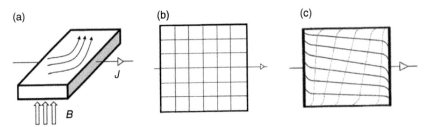

Figure 3.1 The origin of the magnetoresistive effect in semiconductors: (a) the influence of the Lorentz force on the current path length; (b) the current paths and equipotential lines in a conducting rectangular plate in the absence of the external field; (c) the same lines after application of the external magnetic field. *Source:* adapted from [8].

and $\mu = 3 \times 10^4$ cm^2/Vs for InAs). As a material for preparing AMR sensors, the thin film of NiFe alloy (permalloy) is most frequently used. The thin film permalloy exhibits a magnetoresistivity coefficient equal to about 2.5%. The thin film in comparison to the bulk material offers several advantages. The magnetization process of thin film is relatively simple and fast approximate to a single domain model. For full magnetization, a relatively small (about 1 mT), external magnetic field is needed. Thin film in the form of the path enables to obtain large resistance of the magnetoresistor. It is convenient that these films are fabricated using standard semiconductor technology.

The AMR effect is described as a change in the scattering cross-section of atomic orbitals distorted by the magnetic field as illustrated in Figure 3.2. The resistance produced by scattering is maximum when the magnetization direction is parallel (i.e. 0° or 180°) to the current direction and minimum when

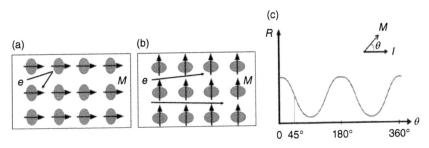

Figure 3.2 Illustration of AMR effect showing distortion of electron orbitals and resulting difference in scattering when the magnetization is (a) parallel to the current or (b) perpendicular to the current direction. (c) Variation of resistance as a function of angle between the current and magnetization. The optimum operation point is at 45°. *Source:* adapted from [9].

the magnetization is perpendicular to the current. In general, the resistance is given as a function of the angle, θ, between the magnetization and current:

$$R = R_0 + \Delta R \cos^2 \theta \tag{3.3}$$

where R_0 is the resistance of the material in the magnetic field $B = 0$, ΔR is the amplitude of the AMR. This function, plotted in Figure 3.2c, shows that the maximum sensitivity and linearity is achieved when the magnetization is at 45° with respect to the current. The 45° alignment is commonly achieved by patterning diagonal stripes of highly conductive metal onto the more resistive AMR material as shown in Figure 3.3a. The current will then run perpendicular to these "barber pole" stripes while the magnetization vector remains preferentially along the long direction of the MR device. The application of an external magnetic field will rotate the magnetization with a resulting change in resistance as shown in Figure 3.3b. The MR ratio for AMR materials is typically a few percent. Integrated AMR sensors are commercially available [9].

If $\rho_\parallel$ is designated as the resistivity of a magnetized and saturated FM material in the direction of the magnetization, and $\rho_\perp$ is designated as the resistivity in the direction orthogonal to the magnetization, it has been shown [3] that the resistivity ρ_0 of such a material, if fully demagnetized, is

$$\rho_0 = \frac{1}{3}\rho_\parallel + \frac{2}{3}\rho_\perp \tag{3.4}$$

From these values, a so-called AMR coefficient or ratio (AMR) is defined as

$$\left(\frac{\Delta\rho}{\rho_0}\right)_{AMR} = AMR = \frac{\rho_\parallel - \rho_\perp}{\frac{1}{3}\rho_\parallel + \frac{2}{3}\rho_\perp} \tag{3.5}$$

Some authors use a different definition of the AMR coefficient, such as $AMR = \Delta\rho/\rho_\parallel$, or $AMR = \Delta\rho/\rho_\perp$, so attention has to be paid to the proper interpretation of a particular text.

(a)   (b)

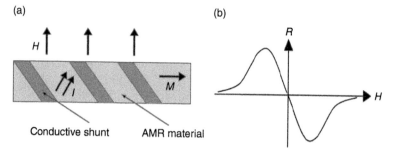

Conductive shunt   AMR material

Figure 3.3 (a) Barber-pole structure of conductive shunts that constrain the current to run at 45° to the rest position for the magnetization. (b) Resistance versus field for a property-biased AMR device. *Source:* adapted from [9].

3.2.1.2 AMR Device Material

There are four chemical elements that exhibit FM behavior above $0\,°C$: Fe ($771\,°C$), Ni ($358\,°C$), Co ($1127\,°C$), and Gd ($16\,°C$). In parentheses, the Curie temperature of these metals is indicated, where the transition from FM to paramagnetic state occurs. Various alloys of Ni, Co, and Fe are common in MR sensors technology. It can be noticed that these elements belong to the group of the so-called transition metals, i.e. they do not have fully occupied 3d-orbitals. Although a phenomenological description of the extraordinary magnetoresistance is possible in rather a noncomplicated way, the origin of this effect cannot be explained without applying quantum mechanical principles on the conductivity in transition metals. The AMR coefficient reaches units of percent in the FM materials, see Table 3.1. In Table 3.1, additional material specific parameters important in the design of AMR sensors are indicated, i.e. characteristic field of anisotropy, magnitude of magnetization in saturation, and magnetostriction coefficient.

Before GMR, there was an active pursuit of devices utilizing AMR [3]. Miller et al. introduced the use of AMR technology in the form of a ring sensor [10]. In the ring approach, a single-layer, current-in-plane $Ni_{80}Fe_{20}$ ring sensing element is fabricated whereby the AMR material is modulated by the radial fringing field from a single magnetic microbead. The ring sensor has outer and inner diameters of 5 and $3.2\,\mu m$, respectively. When the bead is centered over the ring, the radial fringing field rotates the magnetization from

Table 3.1 Parameter of the ferromagnetic material @ room temperature [7].

Alloy composition	$\Delta\rho/\rho_0$ (magnetoresistive coefficient)	ρ_0 (resistivity of demagnetized material)	H_f (characteristic field)	M_s (saturated magnetization)
(%)	(%)	($\times 10^{-8}\,\Omega\cdot m$)	($A\cdot m^{-1}$)	($\times 10^{-5}\,A\cdot m^{-1}$)
NiFe 81 : 19	2.2	22	250	8.7
NiFe 86 : 14	3	15	200	7.6
NiCo 70 : 30	3.8	26	2500	7.9
NiCo 50 : 50	2.2	24	2500	10
NiFeCo 60 : 10 : 30	3.2	18	1900	10.3
NiFeCo 74 : 10 : 16	2.8	23	1000	10.1
NiFeMo 87 : 8 : 5	0.7	72	490	5.1
CoFeB 65 : 15 : 20	0.07	86	2000	1.03

circumferential towards a radial outward direction. This rotation causes a magnetoresistance decrease and a measurable voltage signal in the Wheatstone bridge. Recently, AMR nanostructures with different shapes such as L-shape [4, 5, 11], square ring [12], and zigzag [13–16] have been employed for detection of magnetic beads.

3.2.2 Giant Magnetoresistance

3.2.2.1 Giant Magnetoresistance Effect and Principles

The GMR effect happens in a multilayer structure in which magnetic and non-magnetic thin films are deposited alternately. When there is no external magnetic field, magnetization of the all FM layers are coupled to their neighbors oppositely. At this state, electron spins with both orientations are in low conductance spin channels due to the spin collision at the interfaces between FM and nonmagnetic layers. This state is called the high resistance state (Figure 3.4b). When we apply an external magnetic field, all the FM layer's magnetization is saturated in the field direction. At this state, the electron spins with the orientation in the same direction as the external magnetic field do not have spin collisions at the interfaces, thus they are in a high conductance spin channel. This state is called the low resistance state (Figure 3.4a and c).

The GMR effect was first described independently by Baibich et al. and Binasch et al. through the study of magnetic properties of magnetic and nonmagnetic metal thin film multilayers [17, 18]. It should be noted that the GMR ratio increases when the thickness of the nonmagnetic layers is reduced. Thinner nonmagnetic layer gives larger GMR ratio, but requires a larger saturation field to switch. The magnetoresistance curve is measured at 4.2 K and the maximum magnetoresistance ratio is about 80% with respect to the low resistance state.

Structures exhibiting the GMR effect always take a form of sandwiches of thin films of FM materials separated with a thin [19] conductive layer, so-called

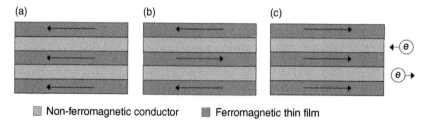

■ Non-ferromagnetic conductor ■ Ferromagnetic thin film

Figure 3.4 Schematic diagram of a current in plane GMR structure. (a) and (c) magnetization of all ferromagnetic layers are saturated in one direction by an external field (low resistance states); (b) magnetization of all ferromagnetic layers is coupled oppositely (high resistance state).

spacer. The resistivity of such a structure, measured usually in-plane for technological reasons, is dependent on the mutual angle of magnetization vectors in the FM layers, being the highest for antiparallel alignment of magnetization vectors ($\rho_{\uparrow\downarrow}$) and the lowest for their parallel alignment ($\rho_{\uparrow\uparrow}$). The GMR coefficient, being of 10–20% at room temperature, is usually defined as

$$\left(\frac{\Delta\rho}{\rho}\right)_{GMR} = GMR = \frac{\rho_{\uparrow\downarrow} - \rho_{\uparrow\uparrow}}{\rho_{\uparrow\uparrow}} \tag{3.6}$$

Let ξ be the mutual angle of magnetization vectors in the adjacent FM layers. The resistivity ρ_{GMR} of the structure is given as

$$\rho_{GMR}(\xi) = \rho_0 - R\frac{\Delta\rho}{2}\cos\xi \tag{3.7}$$

Here $\Delta\rho = \rho_{\uparrow\downarrow} - \rho_{\uparrow\uparrow}$; $\rho_0 = (\rho_{\uparrow\downarrow} - \rho_{\uparrow\uparrow})/\rho_{\uparrow\uparrow}$ (in fact, the resistivity of the structure with orthogonally oriented magnetization vectors).

It has to be emphasized that the resistance of a GMR structure is dependent on the angle of total magnetization vector M in the FM layers only, regardless to the direction of flowing current. This is a significant difference of GMR structures to AMR elements. A modified version of the multilayer device uses only two magnetic layers as shown in Figure 3.5 [9]. The bottom layer is deposited directly on top of an anti-FM "pinning" layer. The anti-FM layer has no net magnetization of its own, but tends to hold the magnetization of the adjacent FM layer fixed in direction. The other layer is free to rotate its magnetization in response to a field.

This structure has been termed a "spin valve" as one can imagine the magnetic field turning the upper layer like a faucet valve to control the flow of spin-polarized electrons through the device. In a properly biased spin valve the rest position of the free layer is made to be perpendicular to the pinned layer so that maximum sensitivity and signal swing is achieved. The response to a magnetic field applied in the direction of the pinned layer is linear over a fairly broad range. The resulting R vs. H function is odd and passes through zero.

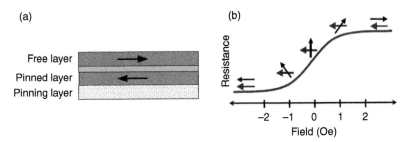

Figure 3.5 (a) Structure of a spin valve (free, pinned and pinning layers). (b) Variation of resistance as a function of the applied field. *Source:* adapted from [9].

3.2.2.2 Mechanism of GMR Effect

The mechanism responsible for the GMR effect is believed to be different from the mechanism responsible for AMR effects [20]. The GMR effect is believed to arise from the spin-dependent scattering, i.e. the interaction of the four electrons of the nonmagnetic metals and the three electrons of the FM metals at the interface between the FM and nonmagnetic layers. The probability of scattering conduction electrons depends on the relationship between the spin direction of the four conduction electrons in the nonmagnetic layer and the direction of the magnetization of the three electrons in the magnetic layers [21]. A Co/Cu multilayer consisting of alternating layers of FM Co and nonmagnetic Cu has been used as in Figure 3.6 [20]. The direction of the magnetic moments of three electrons in Co (grey arrows with large arrowheads) and the four conduction electrons in Cu (small black arrows in open circles) are labeled with the corresponding arrows.

The long black arrows are connected to the conduction electron's scattering path. Under normal conditions, when no magnetic field is applied, such as the case in Figure 3.6a, FM moments in the adjacent Co layers are aligned antiparallel. In this situation, four conduction electrons of Cu with spin up and parallel to the magnetic moments of three electrons of Co pass through the first Co layer without scattering. However, when they encounter the second Co layer that has magnetic moments of three electrons aligned antiparallel to the spin direction of the four conduction electrons of Cu, they are scattered (it is assumed that the spin orientation of the conduction electrons is not changed even after scattering). Similarly, the four conduction electrons with spin down and antiparallel to the magnetic moments of three electrons undergo multiple scattering in each layer that they travel. The movement of both the spin-up and spin-down conduction electrons is therefore, repeatedly interrupted by scattering processes. A repeated scattering, therefore, results in high resistivity. If the magnetic field is applied to overcome the anti-FM coupling (antiparallel alignment of magnetic

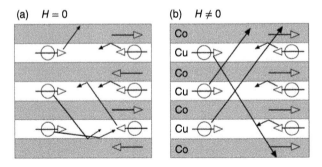

Figure 3.6 Cross-sectional view of a Co/Cu multilayer and the scattering of the four-conduction electrons by the local magnetic moments: (a) at the zero magnetic field and (b) at the applied magnetic field, *H*. *Source:* adapted from [20].

Figure 3.7 (a) Isotropic multilayer (i.e. magnetic moments are randomly orientated) and (b) anisotropic multilayer (i.e. magnetic moments are uniaxially oriented). *Source:* adapted from [20].

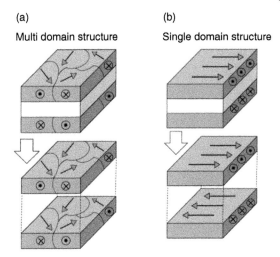

(a)

Multi domain structure

(b)

Single domain structure

moments) and achieve parallel alignment of magnetic moments in the adjacent FM layers as shown in Figure 3.6b, although the spin-down electrons of Cu are still scattered by the magnetic moments of the three electrons of Co, the spin-dependent scattering of the spin-up conduction electrons dramatically reduces. The result is the large decrease in resistivity. The decrease in resistivity is so high that it can go up to two orders of magnitude. For practical multilayers, not only are the spin arrangements along the cross-section of the multilayer important but so are the spin arrangements along the surface of the film.

Figure 3.7 shows (a) isotropic and (b) anisotropic multilayers showing magnetic moments along the surface of the FM layers. In order to have a large GMR effect, a method is required to change the relative orientation of the magnetic moments in the adjacent magnetic layers from a random orientation, as shown in Figure 3.7a to a uniaxial orientation shown in Figure 3.7b. In addition, to have a large GMR effect, the thickness of the nonmagnetic layer must always be less than the mean free path of the electrons. The mechanism causing the GMR effect in granular alloys is the same as that of causing the GMR effect in multilayers. In both the cases, the GMR effect arises due to spin-dependent scattering. For magnetic multilayers, the scattering takes place at the interface of the FM and nonmagnetic layers (Figure 3.6), whereas in granular alloys the same occurs at the interface of the FM and nonmagnetic nanoparticles (Figure 3.8).

Figure 3.8 shows a scenario of spin-dependent scattering between superparamagnetic (circles) and FM nanoparticles (ellipsoids) embedded in a nonmagnetic matrix [22]. The arrows attached to each superparamagnetic nanoparticle indicate random orientation and fluctuation of superparamagnetic nanoparticles. The magnetic moments of the FM particles have orientations determined by their shapes. The arrow connecting superparamagnetic and FM particles

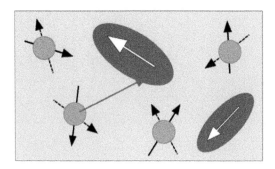

Figure 3.8 The GMR mechanism in the granular alloys consisting of superparamagnetic (circles) and ferromagnetic (ellipsoids) nanoparticles in the nonmagnetic matrix. *Source:* adapted from [22].

shows a path of a conduction electron in the alloy. An electron polarized by the superparamagnetic nanoparticles can undergo multiple spin-dependent scattering due to the orientation of magnetic moments of the FM particles. This results in the GMR effect.

The degree of alignment of the magnetic moments at which spin-dependent scattering takes place determines the GMR effect. The GMR effect is, therefore, influenced by the overall magnetic states in the alloys. The volume concentration of the magnetic particles, their size distribution, the interparticle separations, etc., contributes to the interfacial spin-dependent scattering of the conduction electrons, and, thus, GMR effect. Further details are available in [23].

3.2.2.3 GMR Effect in Multilayers

With the discovery of the GMR effect in magnetic multilayers, search for new materials continued and extended to artificial materials consisting of nanometer-size FM metal granules embedded in nonmagnetic media. This led to the discovery of the GMR effect in Co—Cu [24] and Co—Ag [25] granular alloys, which developed significant interest in further development of GMR granular materials. Further study in magnetic granular alloys is prompted by the simplicity in growth and use of relatively cheap and fast processes [26]. Since then, a large number of magnetic alloys have been reported, e.g. Co—Au [27], FeCo—Au [28], Co—Ag [29], Co—Cu [30], Fe—Cu [31], and Co—Pt [32, 33]. Further works on granular alloys are available in the literature [27, 34–36]. The GMR effect and the field sensitivity (MR ratio per applied field) of granular alloys are much larger as compared to the Hall effect in FM metals and the AMR effect in Ni—Fe alloys [20]. Although the GMR effect in granular alloys is comparable to or larger than the GMR effect in multilayer system, the field sensitivity of granular materials is still low comparatively. However, granular alloys display a number of very interesting magnetic and transport properties such as super-paramagnetism and magnetothermal conductivity, making them attractive for both scientific investigation and application in industry. Table 3.2 lists the GMR effect in Co—Cu, Co—Ag, Co—Au, and Fe—Cu granular alloys

Table 3.2 Reports of the GMR effect in Co—Cu, Co—Ag, Co—Au, and Fe—Cu granular alloys [20].

Alloy	Composition (at %)	Preparation method	Temperature (K)	Field (kOe)	MR (%)	Reference
CO—Cu	$Co_{19}Cu_{81}$	Sputtering	10	20	22	[24]
	$Co_{20}Cu_{80}$	Sputtering	5	20	17	[37]
	$Co_{16}Cu_{84}$	Electrodeposition	300	11	6.2	[38]
CO—Ag	$Co_{70}Ag_{30}$	Pulsed-current	300	10	9.1	[29]
	$Co_{35}Au_{65}$	Pulsed-current	300	1	4.6	[27]
CO—Au	$Co_{15}Au_{85}$	Arc melting	5	20	28	[39]
Fe—Cu	$Fe_{30}Cu_{70}$	Sputtering	5	20	9.0	[36]
	$Fe_{70}Cu_{30}$	Mechanical alloying	300	4	1.5	[40]

prepared using a wide range of methods. A random comparison of MR ratios in these alloys is made among sputtered, melt-spun, mechanical alloying, and electrochemically deposited granular alloys. Among all of the reported results, the overall MR ratio of sputtered Fe—Cu and pulsed-current deposited Co—Ag showed the largest room temperature GMR effect of up to 9.1%.

According to Table 3.2, the deposition method of pulsed-current seems to be one of the best methods when it comes to growing magnetic alloys and nanoparticles. Composition modulated alloys prepared by the pulsed-current deposition method show GMR values that were comparable to the GMR values reported for multilayers and granular alloys [41].

3.2.3 Magnetic Tunnel Junctions

The discovery of GMR in multilayered FM films separated by thin metallic spacers has initiated an enormous research interest, particularly for a wealth of potential applications, e.g. in data storage devices. Fueled by these developments and earlier efforts in tunneling devices [42–45] have discovered that the tunneling current between two FM films separated by a thin oxide layer strongly depends on an external magnetic field, an effect now known as TMR. Since then, the impact of magnetic tunnel junctions (MTJs) on the field of spintronics has hugely expanded, particularly due to the enormous magnitude of the observed magnetoresistances at room temperature and its impact on potential applications [46]. TMR structures are similar to spin valves except that they utilize an ultrathin insulating layer to separate two magnetic layers rather than a conductor (see Figure 3.9). The materials of the insulating layer could be Al_2O_3 [44], Ga_2O_3 [47], MgO [48, 49], and graphene [50]. MTJs are the most sensitive

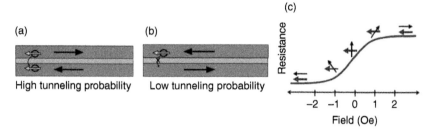

Figure 3.9 Illustration of spin dependent tunneling. When the layer magnetizations are aligned (a) electrons from the top layer can find many available states in the bottom layer to tunnel into. When the magnetizations are opposite (b) then the majority electrons in the top layer cannot tunnel into the bottom layer. (c) The resistance response is similar to GMR spin valves but larger in amplitude. *Source:* adapted from [9].

magnetoresistance sensors with an MR ratio of 20–50%, or as reported, over 200% when using a MgO tunnel barrier [51]. Instead of a Cu spacer like that found in the spin-valve construction, MTJs have a thin insulating layer ($\leq$2 nm thick), which acts as a tunnel barrier. The thickness of the insulating layer can be varied to effectively "tune" the device sensitivity. In addition, the sensing current is directed perpendicular to the relatively large area MTJ layers rather than in the plane of the sensor as used in GMR structures. As one of the distinguished applications of TMR devices, several groups have demonstrated the capability of these devices to detect micro- to nano-sized magnetic beads [52–54]. Production of a 10^6 sensor array, with the promise of each sensor detecting a single magnetic label attached to a single DNA fragment has been pursued and investigated by Wang et al. [54].

When electrons are tunneling between two FM metals, the magnitude of the tunneling current depends on the relative orientation of the magnetization of both electrodes [55]. This can be understood from a few elementary arguments: (i) the tunneling current is, in first order, proportional to the product of the electrode density of states (DOS) at the Fermi level; (ii) in FM materials, the ground-state energy bands in the vicinity of the Fermi level are shifted in energy, yielding separate majority and minority bands for electrons with opposite spins; and (iii) assuming spin conservation for the tunneling electrons, there are two parallel currents of spin-up and spin-down character [55]. As a result of these aspects, the current between electrodes with the same magnetization direction should be higher than those with opposite magnetization. TMR devices are operated in the following current perpendicular to planes (CPP) configuration with contacts on top and bottom of the film stack. The crossing probability is higher when both magnetic moments are aligned in parallel and lower when both magnetic moments are not aligned in parallel [56]. The equation describing the output of these structures is

$$\Delta V = \frac{1}{2} TMRi \frac{R \times A}{Wh} \cos \left(\theta_p - \theta_f \right) \tag{3.8}$$

where TMR is the maximum MR level; i is the biasing current; $R \times A$ is the resistance per area parameter; Wh are the dimensions; θ_p and θ_f are the angle of the magnetization angle of the pinned and free layers, respectively.

3.2.3.1 TMR Structures

In order to guide the reader through the contributions highlighted here, Figure 3.10 shows a number of engineered tunneling structures [55]. The first experiments on spin tunneling has been reported in [42]. In their case, only one electrode is FM (Ni), the other being a superconductor (Al). They have found that though minority electrons dominate the DOS at the Fermi level of Ni, majority electrons are most efficiently tunneling through the thin Al_2O_3 barrier. Later, it is suggested by [57, 58] that, although the dominant species of electrons at the Fermi level of transition metal ferromagnets are minority d-electrons, they do not couple well with the states over the barrier. Instead, highly dispersive majority s-like electrons have a much larger overlap integral with states in the barrier, which leads to a larger transmission probability for these electrons. After these basic papers on FM tunneling, including the first prediction of a TMR effect by [43], it took around two decades to do the same experiment with two FM electrodes, as mentioned in the introduction [44, 45]. It should be noted that in all these experiments Al_2O_3 is preferred as barrier material, primarily since it allows an easy growth of a pinhole-free thin barrier by natural, thermal, or plasma oxidation of Al thin films. Figure 3.10a shows the layout of such an MTJ.

Theory has provided vital evidences that the interface between the barrier and the ferromagnet, and the relevant chemistry or bonding at such an

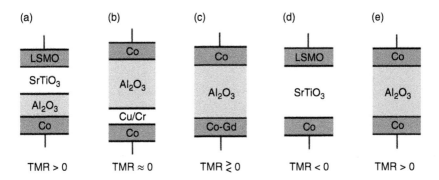

Figure 3.10 (a–e) A number of basic configurations showing, for example, the role of the barrier/electrode combination in relation to the sign and magnitude of the TMR effect; these cartoons are used throughout this paper; see the text. *Source:* adapted from [55].

interface, is crucial for spin tunneling. For example, using first-principles calculations, [59] predicted a sign change for the spin polarization of tunneling electrons depending on where oxygen atoms sit on a Co surface. Since TMR is directly related to the tunneling spin polarization (P) induced by the FM DOS, one may imagine that P is not constant over the whole Fermi surface, and varies depending on which direction in k-space one probes, that is, on the crystallographic orientation of the electrode at the interface with the tunnel barrier. The demonstration of such crystal anisotropy of the TMR is given by [60], who have shown that the use of single-crystalline Fe electrodes of different orientations in MTJs resulted in a substantially different TMR. Inserting an additional layer at the barrier–ferromagnet interface (see Figure 3.10b) has been investigated to rigorously probe the origin of tunneling spin polarization P. [61] show that inserting one monolayer of Cu between the bottom Co electrode and the Al_2O_3 barrier leads to a strong reduction of TMR. Moreover, while the TMR exponentially decays with a length scale of 2.6 Å for a Cu layer, a similar layer of anti-FM Cr induces an even faster exponential decay on a length scale of 1.2 Å [62].

Authors in [63] have further developed these experiments by achieving sharp interfaces between single crystalline Co(001) and Cu(001) using molecular beam epitaxy. They explain that majority electrons tunneling from NiFe into Co would transmit easily as compared to minority electrons, which have a higher probability to be reflected at the Co–Cu interface.

Although most ferromagnets display a positive P in conjunction with Al_2O_3, [64] have reported that Co–Gd alloys can exhibit both positive and negative P systematically depending on the alloy composition. It is known that in these alloys the Co and Gd FM subnetwork magnetization is aligned antiparallel with respect to each other, which may significantly influence the tunneling spin polarization. Now the sign of P depends on the orientation of the respective subnetwork magnetization with respect to the applied field. The P from either of these subnetworks will be positive when its magnetization is aligned with the applied magnetic field, in contrast to the moments of the other subnetwork. Kaiser et al. [64] found that the measured P is the sum of independent spin-polarized tunneling currents from the Co and Gd subnetworks, resulting in a sign change of P with alloy composition. When combined with traditional FM materials with positive P in an MTJ, this leads to positive or negative TMR, depending on the sign of the Co–Gd polarization (see Figure 3.10c).

Chemical bonding at the interface has been predicted to have a great influence on P. Such bonding would influence the tunneling matrix element occurring in Fermi's golden rule which couples initial and final state wave functions, depending on symmetry and overlap arguments. Consider the case of Co–Pt alloys studied by [65]. They observe that the measured P does not change after alloying FM Co with up to 40 at% of nonmagnetic Pt, while the magnetic moment of the alloy reduces by approximately 40% of its initial value for Co.

Arguably, the most decisive experiments demonstrating the relevance of interface bonding effects are those of [66, 67]. In the latter case, MTJs with Co/I/La$_{2/3}$ Sr$_{1/3}$MnO$_3$ (LSMO) were studied, where I could be SrTiO$_3$ (STO), Ce$_{0.69}$La$_{0.31}$O$_{1.845}$ (CLO), or Al$_2$O$_3$. In these experiments, the effective polarization of Co is found to be positive (majority electrons tunnel preferentially) with Al$_2$O$_3$ as barrier, and negative (minority electrons tunnel preferentially) with STO or CLO as barrier. As the P of the STO–LSMO interface is known to be positive, the inverse TMR observed in Co/STO/LSMO junctions is the signature of a negative polarization of the Co–STO interface (see Figure 3.10d). This inversion of the sign of P for the Co–STO interface with respect to the P of Co–Al$_2$O$_3$ is confirmed by growing Co/Al$_2$O$_3$/STO/LSMO junctions which also reveal a positive P for the Co–Al$_2$O$_3$ interface (Figure 3.10e). The negative P of Co when the barrier is STO or CLO can be viewed as a preferential selection of electrons of d-character at the Co–STO and Co–CLO interfaces, as compared to the positive P in Co–Al$_2$O$_3$ where the selection of electrons with s character occurs at the interface. This negative P of the Co–STO interface has later been verified from first principles by [68]. These results again show that tunneling spin polarization (and therefore TMR) should be viewed as a property predominantly determined by the barrier–ferromagnet interface and strongly influenced by the chemistry at the interface.

3.3 Classes of Magnetoresistive Sensors

Now that we have described the principles, materials, and the structures of MR devices, we will discuss about classes of these devices focusing on their different applications. The real applications in which the MR devices have demonstrated their capabilities will be described.

3.3.1 General Purpose Magnetometers

The Earth's magnetic field measurement is a traditional application of the magnetic sensing devices. The Earth has a magnetic field, approximately between 25 and 60 µT, which can be detected by MR sensors. These sensors are widely being used in digital compasses for mobile applications in competition with standard Hall solutions [56]. Table 3.3 shows main parameters of some selected MR compasses [56, 69].

The capability to measure magnetic fields concerning their strength and direction is useful for a number of applications, especially when allowing for field measurement in three spatial directions. Development of a three-dimensional GMR sensor utilizing a ferrite flux guide in order to redirect the field perpendicular to the in-plane direction into the sensing plane has been published [70]. In the beginning stood the simulation and experimental

Table 3.3 Detailed parameters of selected commercial magnetoresistance-based compass [56].

General	Technology	AMR		GMR	MTJ
	Company	MEMSIC	Honeywell	Yamaha	Freescale
	Product	MMC314XMR	HMC5883L	YAS529	Mag3110
PKG	PKG	LGA10	LGA16	WLCSP10	DFN10
	Size (mm^2)	3 × 3 × 1	3 × 3 × 0.9	2 × 2 × 1	2 × 2 × 0.85
I/O	Voltage(V)	1.7–3.6	2.7–5.25	2.16–3.6	1.95–3.6
	Working current (mA)	~2	~2	4	>1
	Samples per s @ mA	50 @ 0.55	7.5 @ 0.10	4	10 @ 0.14
	Interface	12C	12C	12C	12C
	Interrupt	—	—	—	Y
Max ratings	Storage temperature (°C)	−55/+125	−40/+125	−50/+125	−40/+125
	Operating temperature (°C)	−40/+85	−30/+85	−40/+95	−40/+85
	Max exposed field (G)	—	10k	2k	1k
Performance	Range (±G)	4	1–8	3	10
	ADC (output bits)	12	12	10	15
	Resolution (mG)	2	2	6(x,y)/12(z)	1
	Offset (±G)	0.2	—	—	0.01
	Accuracy	2	2	5	—
	Linearity (%FS)	1	0.1	—	1
	Hysteresis (%FS)	0.1	0.0025	—	1
	Repeatability (%FS)	—	0.1	—	—
	Sensitivity TC (%/°)	0.11	—	—	0.1
	Offset TC (mG/°)	0.4	—	—	0.1
	Bandwidth	40	75	40	40
	Noise (RMS)	0.6 mG @ 25 Hz	—	—	0.5 mG
Features	On-chip temp sensor	Y	—	Y	Y
	Single chip integration	—	—	Y	—
	Offset removal	Y	Y	—	—
	Self-test	—	Y	—	—
	Others			Three external AD	Oversampling

validation of the response of a setup of two pairs of commercial dual bridge GMR sensors arranged on a cross-shaped printed circuit board (PCB) with a cylindrical flux guide in the center. This prototype with a footprint of around 20×20 mm^2 was developed further as a single bridge device in order to reduce the device dimensions as well as the power consumption and cost incurred by fabrication [71]. Another similar device was investigated by this group, composed of three commercial GMR sensors mounted on a PCB around a cylindrical flux guide, wound with coils in order to provide a modulation signal [72].

3.3.2 MR Sensors in Harsh Environments

Magnetic field measurement at high temperatures and compatibility of MR sensors with high temperature fabrication processes poses a challenge in the context of MR sensors. For the last 20 years, many investigations dealt with the realization of robust GMR sensors with reliable thermal stability in the range of over 200 °C, comprising of spin valve systems with IrMn, PtMn, or NiMn as an afm due to their high blocking temperatures [73–77]. At the Institute of Micro Production Technology (IMPT) a magnetic field sensor has been developed which can operate at temperatures up to 250 °C. The sensor is characterized by a robust and small design and exhibits a relatively high resistance and low energy consumption. The layer stack consists of materials that provide good thermal stability and high sensitivity. As FM materials, CoFe, convincing by its high Curie temperature, and NiFe, characterized by a low coercivity and anisotropy field, are used. Moreover, the pinned layer is designed as a synthetic antiferromagnet, which increases the thermal stability of the pinning and reduces the influence of the pinned layer's magnetic field on the free layer's sensitivity. As an antiferromagnet, NiMn, featuring a high blocking temperature, is chosen [78]. Besides the thermal stability at 250 °C, the investigations revealed a correlation between the strength of the pinning field and the failure rate of the layer stack at higher temperatures [79].

MR sensors are suited for high (~200 °C) and low (no limit) temperature applications. Only at temperatures far above 200 °C will the properties of the MR layers be degraded or destroyed. Due to their small size and thus weight, mechanical shock has no significant impact on MR sensors. Only very high radiation can affect the MR effect [80].

There are several examples for MR sensors in high temperature applications. One of the high volume applications is a wheel speed sensor based on AMR sensor chips. A very important requirement is the extremely high reliability (<0.1 ppm field return rate) despite, e.g. high operating temperatures of up to 195 °C. More than 120 000 000 AMR sensor chips of Sensitec company have been delivered for this application without a single field return caused by the MR technology [80].

Another example for high operating temperatures in combination with high pressure is the use of MR sensors in calipers (open hole tools) for borehole diameter measurement. The caliper is specified for operation up to 175 °C (350 °F) and 140 MPa (20 000 psi); the sensor system is specified for temperatures up to 205 °C. GMR sensors can also be used in high temperature applications: A valve lift sensor developed by Sensitec company can measure the valve movement (lift and rotation) in a fired engine (environmental temperature approximately 150 °C). Currently these sensors are used in motor test stands. The requirements on the sensor performance are quite challenging: a contactless measurement principle with a resolution below 10 μm suited for the measurement of linear speed above 100 m/s. All of these requirements can be met with a GLM Tooth Sensor Module, which detects the deformation of the magnetic field caused by a tooth structure in the valve. The measurement is not even affected by oil contamination. The VLS sensors are used in the development phase of gasoline engines to measure dynamic effects and to detect errors in the valve train. In the very near future these sensors will not only be applied to test-stands but also to racing cars to improve the engine control and thus the performance of the car.

Example for low temperature environments are current sensors for More Electric Aircraft applications. Following on from the Power Optimized Aircraft (POA) and More Open Electrical Technologies (MOET) initiatives, the European Union is now part funding the Clean sky Initiative, where one of the main activities is the replacement or enhancement of hydraulic and pneumatic systems by means of electrically driven actuators, supported by the latest sensor technology [80]. The focus is on the electrification of high-lift systems, the landing gear, and the environmental control system. MR sensors have been applied successfully in demonstrators in all three areas for either angle sensing or current sensing applications [81]. Another aspect in harsh environments is mechanical shock. A good example for MR sensors under high mechanical shock is the use of GMR Tooth Sensors for process control in a Böllhoff riveting tool [80]. The drive unit is equipped with a refilled tooth structure as a passive magnetic scale – an active scale will lose its magnetization during operation due to the high shock. Two Sensitec GLM modules sample this scale and provide the position measurement information. Mechanical shock on the sensor unit up to 1500 g has been measured during operation with resulting forces above 10 kN.

3.3.3 Electrical Current Sensing

Traditionally, electrical current has been measured by means of shunt resistances, coils, and solid-state sensors [82]. The Ohm's law is the basis of the first method and variations of the Faraday's law are applied to the second case. Here the focus will be on third option, where the magnetic field generated by a current flow is detected by a solid-state magnetic sensor. This general scheme can

Table 3.4 Magnetic field generated by an electric current measurement for different geometries [82].

$$\oint H \cdot dl = I$$

Cylindrical wire	Printed circuit board strap
(a)	(b)
$H(r) = \dfrac{I}{2\pi r}$	$H(r) = \dfrac{1}{4\pi}\displaystyle\int_{s} \dfrac{J(r') \times R}{R^2} dr'$
	$R = r - r'; R = |R|$

be applied to the measurement of a current driven by a wire (Table 3.4, left) or by a conductive strap in a PCB or an integrated circuit (Table 3.4, right). AC/DC currents can be measured in this way with small, cheap, and contact-less systems. When dealing with GMR or giant magnetoresistance impedance (GMI) sensors, excellent sensitivities are achieved. For example, in [83], a current sensor based on the GMI characteristics of FM wires is developed. When excited at 100 kHz, sensitivities of about 2.5 V/(V·A) in a 0–0.3 A range are demonstrated.

3.3.3.1 Industrial Electronics Applications (Large to Medium Currents)

In [84], a specific spin valve sensor for industrial applications is designed, characterized, implemented, and tested. The sensor was soldered onto a PCB current track and encapsulated chip-on-board. A full bridge (active in pairs) with crossed axis configuration was utilized. The sensor displayed a linear range up to 10 A. In [85], a novel design principle is presented. The principle of operation

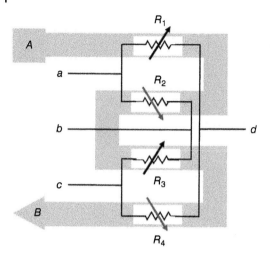

Figure 3.11 Current sensor configurations. *Source:* adapted from [82].

is depicted in Figure 3.11. With this configuration, the current flows from-left-to-right above R_1 and R_3, and from-right-to-left above R_2 and R_4. Consequently, when a current is driven (from A to B), and depending on the sign, resistances R_1 and R_3 increase/decrease their values and resistances R_2 and R_4 decrease/increase their values, thus obtaining a full Wheatstone behavior. The sensor is fed through terminals a and b, and the output is taken between terminals c and d. Due to the particular arrangement of the magnetoresistors, this sensor is theoretically insensitive to the external magnetic field, therefore minimizing the possibility of interferences (below 1% [85]). This particular approach has been successfully applied to PCB-IC mixed technology-based moderate current sensors, with sensitivities close to 1 mV/(V·A).

3.3.3.2 Differential Currents

Differential currents can also been measured with the help of GMR sensors [86]. A GMR sensor (AC004-01, from NVE) is placed into two Helmholtz coils, carrying the currents to be compared. When both currents are identical, the magnetic field in the middle point of the coils is zero, and so the output voltage of the sensor. The system was tested in a house-hold application, demonstrating to be useful for detecting differential currents below 30 mA.

3.3.3.3 Switching Regulators

In [87] the performance of the sensor presented in [84] was compared with a common Hall effect current transducer (LEM, LA 55-PS/P1) within a high-frequency bi-directional three-phase rectifier, to be used in accelerator applications at the European Laboratory for Particle Physics (CERN) [87]. The spin valve sensor displayed excellent figures regarding noisy and heat environment, due to their intrinsic properties.

3.3.3.4 Wattmeter

Separating real-time sampling and A/D conversion of the current $i(t)$ and voltage $v(t)$, followed by fully digital processing of acquired data is the more straightforward scheme for power measurement. This is the approach followed, for example, by Ramirez et al. In [88], an electronic system to measure active, apparent, and reactive energies and power delivered to an AC line load is presented. By using of an appropriate analog electronic multiplier, two signals proportional to the voltage and current, respectively, can be multiplied in real-time, as suggested in Figure 3.12a. Thus, the output of such a transducer is the instantaneous power of the signal defined as:

$$P(t) = i(t) \cdots v(t)$$

Even though power transducers based on Hall sensors as multiplying elements can be used for direct power measurement, their insufficient sensitivity usually results in the need of FM cores to concentrate the magnetic flux into the sensor area. The higher sensitivity of GMR-based sensors makes them as potential substitutes of Hall sensors for this application. The basic idea of using an MR element as an analog multiplier is very simple: the Wheatstone bridge of the MR sensor is supplied by a signal, which is proportional to the voltage of the measured signal. At the same time, current proportional to the current of the measured signal is led through a coil, generating magnetic field which the Wheatstone bridge is exposed to. The output (diagonal) voltage of the bridge is (linearly) dependent on the acting magnetic field, and at the same time, it is linearly dependent on the supplying voltage. As a direct consequence of these two facts, the output is dependent on the multiple of the two signals. A graphical scheme is presented in Figure 3.12b. The idea has been recently applied by using a KMZ51 AMR-based commercial sensor [89]. The substitution of the AMR by a GMR-based sensor is currently under study.

An interesting application of GMR sensors for measuring and controlling electrical power during charging and discharging batteries is presented in [90]. A specific circuit is implemented for taking advantage of the multiplying

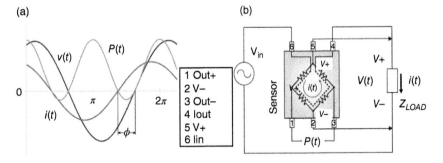

Figure 3.12 Power measurement with a MR sensor. (a) Description of instantaneous power. (b) Possible configuration with a Wheatstone bridge MR sensor [82].

characteristics of a GMR sensor. The system demonstrated its validity in 12 V batteries with charging/discharging currents up to 4 A.

3.3.3.5 IC Current Monitoring

Some work has been reported regarding the application of GMR-based sensors to the electrical current measurement at the IC level. Authors have demonstrated the applicability of spin-valve structures to the measurement of low electric currents [91, 92]. In these works, authors introduced the concept and some fabrication parameters were established. In [93], the potentiality of spin valve based full Wheatstone bridges for low current monitoring at the IC level is demonstrated.

3.3.4 Automotive Applications

GMR-based devices have already entered various automotive applications. They are very well suited for a wide range of different automotive applications, such as angle and speed sensing because of their high sensitivity and low noise [69]. Mass production combined with high reliability even in harsh environment meets all the automotive quality requirements and provides a cost effective solution with high performance. Among the target applications in an automotive environment are:

- Steering angle measurement.
- Rotor position measurement for motor commutation in a brushless-DC motor
- Speed sensing for wheel speed measurement (ABS Sensor).
- Crank shaft speed and position sensing with direction information.

All of these can be classified as speed measurement or angle measurement. A typical GMR angle sensor consists of several GMR resistors arranged in two bridges, one for each orthogonal direction [56]. Meandered geometries are used for increasing the total resistance to the kiloohms (kΩ) range. In this way, they provide a sine and a cosine signal that can be used in the calculation of the absolute angle of the magnetic field vector. Due to this measurement principle, only the field direction, not the field magnitude, is relevant. The use of the GMR principle allows the measurement of angles in the full range of 360° in contrast to AMR-based sensors that cover only 180°. Some of the automotive angle sensing applications are safety relevant. To support such applications, additional features can be implemented to account for the requirements of the ISO26262 covering functional safety aspects [94].

3.3.4.1 BLDC Rotor Position Measurement

In order to ensure exact commutation of electric motors, the position of the rotor has to be measured very accurately over a wide speed range [69]. The position is transferred to the control unit, which generates the necessary

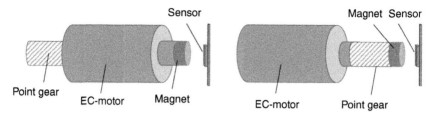

Sensor · Magnet Sensor · Point gear · EC-motor · Magnet · EC-motor · Point gear

Figure 3.13 End of shaft configuration of a diametrically magnetized magnet and GMR angle sensor. *Source:* adapted from [69].

commutation sequence for the motor. For optimized efficiency and lowest torque ripple, field oriented control algorithm is used which needs very accurate information of the rotor position. GMR angle sensors can be used for various motor topologies with different numbers of pole pairs. Figure 3.13 depicts the setup with the GMR angle sensor and a diametrically magnetized magnet.

As GMR angle sensors have a measurement range of 360°, even and odd numbers of pole-pair motors can be used. There are several possibilities of the sensor concept especially whether the raw data or the calculated angle is transmitted to the microcontroller. Which option is preferred depends on the complete system architecture. In case the raw data, i.e. sine and cosine values, are transmitted, the angle calculation has to be done in the microcontroller. This can be performed by the use of the CORDIC algorithm (Coordinate Rotation DIgital Computer), which calculates the arctan of the ratio of the sine and cosine value and thus obtains the absolute angle. Another possibility avoiding the CORDIC algorithm is a tracking loop, which minimizes the angle error calculated out of the measurement data (sine and cosine) and an estimated angle value. Details of this approach can be found in [95].

3.3.4.2 Steering Angle Application

The Steering Angle Measurement is one of the most important criteria for controlling vehicle dynamics. It is the only driver information used by the Electronic Stability Control (ESC) to detect the desired vehicle driving direction [69]. Basically, two different places in a car are used to detect the steering angle:

1) Passenger Compartment: Sensor is located behind the steering wheel; reduced requirements on temperatures ($T_A < 85\,°\text{C}$), vibration and dust.
2) Engine Compartment: Sensor is mounted on the steering shaft near the engine; high robustness against temperature ($T_A > 125\,°\text{C}$), vibration and dust required.

3.3.4.3 Crankshaft Speed and Position Measurement

The main advantage of GMR-based sensors in a speed application is the high sensitivity and a high S/N ratio [69]. This gives on the one hand a large air gap capability which cannot be realized with a silicon-based Hall sensor. On the other hand, the GMR sensing principle has a low noise, which gives a very good jitter performance in the application. This provides superior performance for a crankshaft sensor in engine management. In this application, not only the speed is measured but also a precise knowledge of the crankshaft position is required. Active magnetic sensors in advanced engine management systems provide a digital switching signal, which maps the mechanical teeth profiles or the magnetic domains of a passing pole wheel. Subsequent processing in the microprocessor determines the current speed or angle position of the target wheel from this switching signal. This data is further processed for accurate ignition control and misfire detection. The sensor needs to provide a phase accurate output signal for magnetic input frequencies from 0 to 10 kHz and over an amplitude range of approximately 1–100. The required temperature range is from −40 to 150 °C with an air gap range of 0–2.5 mm.

3.3.4.4 Wheel Speed Measurement for ABS and ESC Systems

Speed sensors are widely used in modern cars especially for safety applications as ABS (antilocking system) and ESC (electronic stability control) to monitor the speed of each wheel [69]. These systems control actively the wheel speed and avoid a blocking of the wheel or instable driving conditions by braking individual wheels to stabilize the car and reduce braking distance. The wheel speed information, however, is used for a large variety of further systems as, for example, parking assistant or hill holder. GMR-based speed sensor can meet requirements with their superior signal-to-noise ratio whereas the use of Hall-based sensors has much more limitations in this application. The GMR-based wheel speed sensor is very similar to the GMR-based crank sensor Main difference is that speed sensors use a two-wire interface as output signal. As wheel speed sensors are mounted directly at each wheel remote from the ECU, this interface can save cost for wiring. The speed signal is transmitted by a modulated supply current, which is detected in the ECU. For sensors without the requirement of providing a direction signal, the current switches between 7 and 14 mA. The frequency of the output pulses corresponds directly to the north/south pole transition of the pole wheel and thus to the wheel speed.

3.3.5 Magnetoresistive Elements in Data Storage Applications

Flexible AMR sensors have been used in the magnetic storage of data on a technical surface [96]. Inspired by classical hard drive magnetic storage technologies, a read/write head was developed in order to obtain a method to store critical product information intrinsically on a component. This will help a manufacturer as well as a user to identify and better apply a component [97]. By analogy with the classical hard drive magnetic storage technologies [98], a write head that generates a magnetic stray field near the air gap in order to magnetize the storage medium as well as a sensor functioning as a read head were developed and tested successfully [98]. As flexible substrate, Kapton® foil was used due to its outstanding physical properties. Subsequent to preliminary works [99], multilayer structures such as the soft magnetic head poles and the MR element have been fabricated on the Kapton polyimide film.

3.3.6 Space

In space sector applications, mass, volume, and power savings are important issues. GMR sensors are excellent candidates not only in planetary magnetometry, but also as magnetic encoders and angular or position sensors [56]. It must be mentioned that space is an environment of extreme parameters, including wide temperature swings, very low pressures, moderate to high radiation, mechanical vibrations and impacts, and so on. GMR sensors have been used on several occasions in different satellite missions. MR sensors are particularly interesting for space applications, because small mass, small volume, high robustness under difficult operating conditions or low power consumption are especially important here [80]. For example, radiation is a huge problem for semiconductor-based sensors (e.g. Hall effect based magnetic sensors) and can cause severe malfunction. The broad temperature range and the mechanical load as well as the high risk of contamination during operation complicate the use of optical sensors. These limitations do not apply to MR sensors. For instance, the environmental conditions during transportation to and operation on Mars and the resulting requirements for the sensors are quite challenging:

- Wide temperature range: −130 to +85 °C
- Solar and cosmic radiation
- Mechanical shock
- Small and lightweight
- Low power consumption
- High reliability

3.4   Modeling and Simulations

3.4.1   Finite Element Modeling and Methodology

There are a great many numerical techniques used in engineering applications for which digital computers are very useful. The numerical techniques generally employ a method which discretizes the continuum of the structural system into a finite collection of points (or nodes)/elements called finite elements. The most popular technique used currently is the finite element method (FEM). There are other methods like the finite difference method (FDM) and the boundary element method (BEM).

3.4.2 Finite Element Method

In the FEM of analysis a complex region defining a continuum is discretized into simple geometric shapes called finite elements. The material properties and the governing relationships are considered over these elements and expressed in terms of unknown values at elements corners. An assembly process duly considering the loading and constraints results in a set of equations. Solution of these equations gives the approximate behavior of the continuum. The basic steps in the FEM are: (i) Discretization of the domain; (ii) Identification of variables; (iii) Choice of approximating functions; (iv) Formation of the element stiffness matrix; (v) Formulation of the overall stiffness matrix; (vi) incorporation of boundary conditions; (vii) Formulation of element load matrix; (viii) Formation of the overall load matrix; and (ix) Solution of simultaneous conditions. The advantages of FEM are: (i) fast, reliable and accurate; (ii) it can analyze any structure with complex loading and boundary conditions; (iii) it can analyze structures with different material properties; (iv) this method is easily amenable to computer programming; and (v) it can analyze structures having variable thickness. Main disadvantages of this method are: (i) the cost involved in the solution of the problem; (ii) it is difficult to model all problems accurately and the results obtained are approximate; (iii) the result depends upon the number of elements used in the analysis; and (iv) data preparation is tedious and time consuming.

3.4.3 Finite Difference Method

In the finite difference approximation of differential equations, the derivatives in the equations are replaced by difference quotients of the values of the dependent variables at discrete mesh points of the domain. After the equations are replaced by difference quotients of the values of the dependent variables at discrete mesh points of the domain. After imposing the appropriate boundary conditions on the structure, the discrete equations are solved obtaining the values of the variables at mesh points. The technique has many disadvantages, including inaccuracies of the derivatives of the approximated solution, difficulties in imposing boundary conditions along curved boundaries, difficulties in accurately representing complex geometric domains, and the inability to utilize nonuniform and nonrectangular meshes.

3.4.4 The Boundary Element Method

The BEM developed more recently than FEM, transforms the governing differential equations and boundary conditions into integral equations, which are converted to contain surface integrals. Because only surface integrals remain, surface elements are used to perform the required integrations. This is the main advantage of BEM over FEM, which require three-dimensional elements throughout the volumetric domain. Boundary elements for a general three-dimensional solid are quadrilateral or triangular surface elements covering the surface area of the component. For two-dimensional and axisymmetric problems, only line elements tracing the outline of the component are necessary. Although BEM offers some modeling advantages over FEM, the latter can analyze more types of engineering applications and is much more firmly entrenched in today's computer-aided-design (CAD) environment. Development of engineering applications of BEM is proceeding however, and more will be seen of the method in the future.

3.4.5 MR Sensors Simulation and Modeling

In recent years, an enormous surge of works has been carried out to develop new methods for detection of a wide range of biomolecular targets in life-science applications, medical diagnostics, and pharmaceutical discovery. Development of high-speed, reliable, accurate and high-resolution biosensing platforms continues to be driven by the huge market potential for biodetection systems. Computer-aided modeling and simulation towards optimal design of biosensing systems has proven their feasible functionality and reliable performance [100]. Simulation enables product comparison in a range of complex physiological environments. Therefore, prototype fabrication cost will be decreased due to optimum design of the material and instrument. TMR device has been used for *Escherichia coli* detection by manipulating superparamagnetic beads to a sensing area [101]. Instead of biochemical immobilization layer, which is commonly used in magnetic biosensor systems, in this work the trapping has been used. Replacing the biochemical immobilization layer by the trapping well greatly simplifies the detection process. A finite-element simulation is performed (using COMSOL software) to calculate BM, the magnetic flux density generated by the current in the conducting line, and BStray, the magnetic flux density of magnetic beads inside the trapping well.

The susceptibility of the superparamagnetic beads used is 0.79 (Dynabeads® M-270). In the model, the conducting line has a width of 6 μm, a length of 15 μm and a thickness of 0.3 μm. TMR sensor surface has a width of 3 μm, a length of 15 μm and zero thickness. The sensor surface is separated from the conducting line by a 200-nm thick layer of SiN and a 60-nm thick nonmagnetic top electrode of the TMR sensor. A direct current of 30 mA is applied to the conducting line in x-direction. The current creates a magnetic field, which changes the resistance of the TMR sensor. By whole device simulation of the bead in the corner of TMR (symmetry does not apply in this case), the average value of y component of B_{Stray} is calculated $1.233e^{-2}$ mT, corresponding to $V_{Stray} = 155.9$ nV. For the bead in the center of TMR, the average value of y component of B_{Stray} is obtained $7.638e^{-3}$ mT, and V_{Stray} is 96.56 nV. As expected, the output voltage of the TMR sensor caused by a single bead is the highest when the bead is located at the center of the trapping well and is the lowest when the bead is located at its corner. The output voltage caused by a single bead should be between 96.56 nV (bead at a corner) and 219.1 nV (bead at the center). This is in good agreement with the experimental results presented by authors in this study. Following previous work, Li and Kosel [102] have demonstrated the capability of TMR biosensor in the detection of *E. coli*. The trap is formed by a current-carrying microwire that attracts the magnetic beads into a sensing space on top of a TMR sensor. For one bead on the chip's surface at the center of the micro-wire ($Y = 0$, $Z = 1.4$ μm) V_{Stray} has been obtained 87.52 nV. This represents the highest value that can be expected from an individual bead. When the TMR sensor is fully covered with magnetic beads, two rows of magnetic beads in the first layer contribute the most to the total signal. When the bacteria are attached to the surface of the beads, a part of the sensing space will be occupied by them, yielding a decrease in the sensor signal. As one of the advantages of this technique and compared with magnetic biosensors previously developed by this group, it does not require surface biofunctionalization and procedure can be completed within half an hour. The same authors have used a bead concentrator, consisting of gold microstructures, at the bottom of a microchannel to attract and move magnetic particles into a trap [103]. In this work, to sense targets, two kinds of solutions have been arranged; one comprising only superparamagnetic particles, the other one containing beads with the protein bovine serum albumin as the target and fluorescent markers. Owing the size difference between bare beads and beads with target, less magnetic beads were immobilized inside the volume chamber in case of magnetic beads with target as compared to bare magnetic beads. A simulation using commercial finite-element software (COMSOL) is performed to calculate the magnetic field and magnetic force. The model is the same to the fabricated device. The height of the gold microstructures is 300 nm. The current applied to the gold microstructure beneath the chamber is 30 mA. The other currents are increased to keep the same current density in all the gold microstructures. The height and width of the chamber made from SU8 is 5 and 8 μm, respectively. The magnetic bead has a diameter of 2.8 μm and a magnetic susceptibility of 0.17, which are the values of a Dynabead M-270. The strongest field exists near the gold microstructures. The field strength rapidly diminishes when moving away from the gold microstructures. The advantage of proposed method is that complex biological treatment of the sensor surface is not required, since a combination of magnetic forces and a mechanical trap is used to immobilize magnetic beads. The magnetic field generated by electrical currents can also be used to magnetize the magnetic beads. This eliminates the need for an external magnetic field source, which is commonly required for MR biosensors. Another aspect worth to mention is that the concentrator would allow this system to operate on droplets rather than using microfluidic channels, which would reduce the complexity.

Micromagnetic simulation for detecting magnetic beads has been also performed by using a high-sensitivity spin-valve sensor as the detector in [104]. The magnetic beads polarized by a DC magnetic field contribute to a nonuniform dipole field, which can affect the magnetization state of spin-valve sensor, leading to a detectable resistance change.

The dipole field created by the magnetic beads is in the reversed direction, which partially cancels the applied field, resulting in a small change of the magnetization of the free layer along the clockwise direction. Micromagnetic simulation study showed that single or several 2 μm diam microbeads can be detected with about 98 μV voltage change per bead, indicating that the single molecule detection with a micron-sized bead marker is possible. Nevertheless, single nano-sized bead as small as 250 nm shows low signal of about 34 μV, which is hardly to be detected by using the present spin-valve sensors.

Li et al. [105] have fabricated a series of highly sensitive spin valve sensors on a micron scale that successfully detected the presence of a single superparamagnetic bead (Dynabeads M-280, 2.8 mμ in diameter) and thus showed suitability for identifying biomolecules labeled by such magnetic beads. By polarizing the magnetic microbead on a spin valve sensor with a DC magnetic field and modulating its magnetization with an orthogonal AC magnetic field, they observed a magnetoresistance (MR) signal reduction caused by the magnetic dipole field from the bead that partially canceled the applied fields to the spin valve. A two-dimensional micromagnetic simulation for the spin valve sensors has been performed using the OOMMF software [106]. They have used OOMMF to simulate only the magnetization behavior of the $Co_{90}Fe_{10}$ free layer to obtain the corresponding MR responses. The saturation magnetization (1540 emu/cm^3) and exchange stiffness (1.53 μerg/cm) of $Co_{90}Fe_{10}$ were obtained from [107]. The uniaxial anisotropy field was found to be 40 Oe, close to the reference value of 32 Oe [107]. The cell size was chosen to be 25 nm for the micron-sized sensors. The Dynabeads are considered as a magnetic dipole, and their susceptibility was experimentally found to be approximately 0.04. The resistance changes $\Delta\widetilde{R}_{sen,simu}$ from simulations are presented in Table 3.5. It can be

Table 3.5 Experimental data for two spin-valve sensors with a single 2.8 mm diameter magnetic bead (Dynabeads M-280) and the micromagnetic simulation results. The voltages are all rms values [105].

Sensor size (μm^2)	Active area (μm^2)	H_t (Oe rms)	H_b (Oe)	V_{bias} (V)	R_2/R_1 (kΩ/kΩ)	V_{sig}^0/V_{noise} (mV/mV)	V_{sig}/V_{noise} (mV/mV)	$\Delta\hat{R}_{sen}$ (mΩ)	$\Delta\hat{R}_{sen,sim}$ (mΩ)
3 × 12	3 × 4.1	32.0	120.0	30.0	9.13/9.53	0.03/0.09	1.2/0.1	5.2	5.6
2.5 × 10	2.5 × 3.8	38.0	94.0	100.0	22.2/24.8	0.04/0.15	3.8/0.3	11.9	13.3

seen that the simulations are consistent with the measurements. Further experiments and showed that these sensors or their variations can detect 1–10 Co nanoparticles with a diameter of about 11 nm, and are suitable for DNA fragment detection.

Li et al. have presented an external field-free magnetic bio sensing structure, which will be useful for magnetic bio sensing system miniaturization [108]. This structure is based on unique patterned grooves embedded in spin valve biosensor and employs the stray fields from the free and pinned layers of biosensor for magnetic nanoparticle (MNP) magnetization. Micromagnetic simulation has been carried out using OOMMF software [109], to simulate the magnetization behavior of the free layer under the stray field from the pinned layer and the dipole field from the MNPs. In the simulation, both the free and pinned layer with a groove structure are divided into small magnetic cells with the same size (5 nm). Each magnetic cell has its own magnetic moment and interacts with all other cells. Before the MNP bonding, the effective field on the free layer is the sum of the stay field from all the magnetic cells of the pinned layer. The MNP is magnetized by the total field from all the magnetic cells of the free and pinned layers. The dipole field from the MNP is discretized and incorporated into the OOMMF input file as well as the stay field from the pinned layer on the free layer. The averaged magnetization orientations of the free layer are computed from the magnetization distribution of the magnetic cells by OOMMF. Results showed a maximum signal to noise ratio of 18.6 dB from one 8 nm radius iron oxide MNP locating in the center of the groove structure and the signal strength increased with the MNP position near the groove corner. In another work, Wirix-Speetjens et al. have described a detection system based on a magnetic spin-valve sensor that is capable of giving position-time information of the magnetic behavior of one single bead [110]. The results obtained with this system for the detection of a single particle signature are then compared with simulations. For this comparison, they have developed a model where an additional particle–substrate separation distance is included. This distance is determined by a force balance of the perpendicular forces acting on the magnetic particle, including the magnetic and electrostatic force. These simulations agree well with the single particle detection experiment. A finite element analysis using Maxwell 3D from Ansoft has been applied for the simulations. This tool requires the value of the relative permeability of the particle's material. This relative permeability is obtained by measuring the magnetic susceptibility, which is related to it, using a SQUID of quantum design. From the SQUID data, a value of $\chi = 0.257 \pm 0.015$ for the susceptibility, and incorporating the particle's demagnetization field, a value of 1.28 ± 0.02 for the relative permeability has been obtained [111]. The model illustrated the importance of the additional particle/substrate separation distance, which was found to be 1250 nm for their particle detection system. When this separation distance is not taken into account, the peak-to-peak signal increases from 250 to about 800 µV. Authors have concluded that this additional separation distance is an important parameter and needs to be taken into account for modeling the sensor response of unbound particles.

3.5 Design and Fabrication Technologies

3.5.1 GMR Devices

GMR sensors fabrication involves deposition, patterning, and encapsulation steps, different from those typically used for semiconductor industry. In contrast to semiconductor processing, GMR device fabrication is based on low temperature techniques; therefore doping-related processes such as diffusion or implantation are usually not required [69]. Materials associated to the fabrication of GMR sensors and devices slightly differ from those used in standard Bi-CMOS processes. Regarding the devices holder, GMR can be deposited on silicon wafers but glass, sapphire or flexible substrates can also be used. In GMR device fabrication, the fabrication of magnetic layers requires the use of additional magnetic materials (Iron, Cobalt, Nickel, Manganese, and their alloys), different metals (e.g. Copper, Ruthenium), and additional oxides (Al_2O_3, MgO, ...). Each of these materials has particular requirements in terms of deposition technology and conditions or system contamination that need to be specifically considered and optimized.

3.5.1.1 Deposition Techniques

GMR structures are composed by multilayered engineered structures [112] based on nanometric to sub-nanometric thick layers of FM materials (e.g. Co, CoFe, NiFe) separated by a nonmagnetic spacer (Cu). Adequate deposition techniques, namely those using ultrahigh vacuum systems and providing a thorough control of the thickness of the deposited layers, are essential for the proper functionality of so obtained devices.

3.5.1.1.1 Physical Vapor Deposition Sputtering

One of the more common technique used for depositing thin films onto substrates is cathodic sputtering [113]. The sputtering process occurs when an accelerated ion collides with a solid target material. When the ion kinetic energy is sufficient ejection of an atom from the matrix takes place due to the momentum transfer. Figure 3.14 displays a sketch of sputtering's basic principle. The thin film deposition requires pumping the reaction chamber to a low enough pressure, so that the water and oxygen adsorbed at the chamber walls is reduced. The deposition requires inert gas (Argon or Xenon) in order to produce the plasma, typically at few mTorr. A high voltage is then applied to the target holder producing an electrical discharge that allows the ionization of the gas and hence leads to the plasma. The produced ions are then attracted toward the cathode hitting the target.

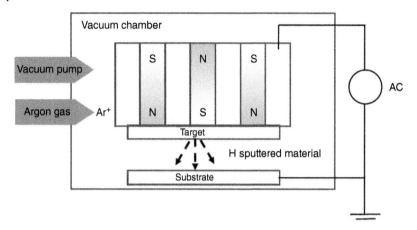

Figure 3.14 Basic schematics of a magnetron sputtering process and sketch.
Source: adapted from [69].

The ions with energy above the threshold can extract atoms from the target material. These atoms are deposited onto the substrate, usually facing the target, and thus forming a layer of material. During the collision process, some secondary electrons are also produced, promoting a sustainable plasma at lower pressures. Furthermore, a magnetron can also be placed near the target to increase the ionization yield (magnetron sputtering). In particular cases, the ionization process can also be assisted by means of thermo-emitted electrons from a filament.

Ion Beam Sputter Deposition   Although not as widely used as magnetron sputtering, the Ion Beam Deposition (IBD) system provides a good film thickness uniformity and higher deposition control due to the low deposition rates employed, enabling also epitaxial growth under particular conditions [114, 115] and higher deposition textures [116, 117].

Figure 3.15 shows an example of a basic IBD schematics in "Z" configuration [118]. Although the physical principle is similar to sputtering, in this case, the plasma is created and confined in an ion gun being then accelerated towards the target through voltage applied into a grid set (graphite or tungsten). RF or DC Kauffman ion sources are usually used, where the ionization process is confined (Ar or Xe), allowing also reactive depositions and in-situ reactions (Nitrogen or Oxygen). The size of commercial sources typically varies from 2.5 to 30 cm and the ion energy is about a few to several hundreds of eV [119]. Furthermore, the basic configuration of a typical IBD system normally includes an assist gun, used either for assisted deposition or ion-milling etching [120, 121]. An automatically interchangeable target holder (four to eight targets) can be used in GMR multilayer deposition without vacuum break, with deposition rates below 1 nm/s.

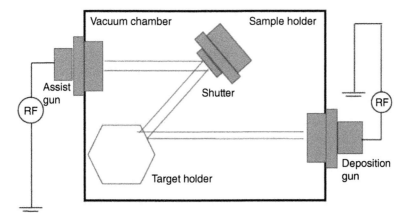

Figure 3.15 An example of a Ion Beam Deposition (IBD) basic schematics in *Z* configuration. *Source:* adapted from [69].

3.5.1.1.2 *Chemical Vapor Deposition*

The deposition of thin films by means of Chemical Vapor Deposition (CVD) is based on the decomposition and/or reaction of different gaseous compounds. With this technique, the considered material is directly deposited onto the substrate surface from a gas phase [122]. High fusion temperature materials such as poly-silicon, silicon oxide, or even heavy metals can be deposited by CVD, with excellent stoichiometry when compared with other deposition techniques. Regarding GMR sensors fabrication, CVD is mainly used in the deposition of insulating layers (silicon oxide or silicon nitride) leading to good quality layers with moderate cost equipment. This method can be shared with semiconductor processing. A schematic of a basic CVD system is depicted in Figure 3.16. It is composed by a heating system and a reactor. The reactor includes the substrates holders and the required gas inlets, depending of the particular material to be deposited. A gas extraction outlet, together with a vacuum pump, is also included.

3.5.1.1.3 *Electrodeposition*

Electrodeposition usually refers to deposition of a metal or an alloy from an electrolyte by passing a charge between the two electrodes located in the electrolyte [123–125]. This is a widely implemented method, either in industry or research, and enables a control of the length of the nanostructure by the duration of the electrochemical process. Besides, this technique is also not expensive, versatile, and does not require vacuum equipment. Several parameters influence the kinetics of the electrodeposition reaction, namely: (i) electrolyte temperature affects the ions diffusion velocity and the diffusion of already reduced atoms on the substrate surface; (ii) stirring favors the electron diffusion, enables

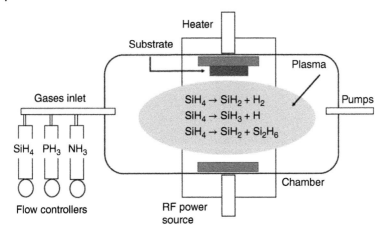

Figure 3.16 A schematic of a basic Chemical Vapor Deposition (CVD) system, composed by a heating system and a reactor. *Source:* adapted from [69].

the removal of H_2 gas which can inhibit the process, keeping the electrolyte concentration and the electrolyte/substrate interface pH constant; (iii) the electrodeposition potential determines the species and the corresponding quantity that will be deposited onto the substrate; (iv) finally the electrolyte composition. Usually, for metallic and magnetic single element nanostructures a standard Watts bath can be used [123]; however, for multilayers deposition from a single bath, all species need to be present in the electrolyte.

3.5.1.2 Patterning

The silicon-based semiconductor industry nowadays relies mainly in ultraviolet lithography of hard or software-designed masks, combined with physical/chemical etching processes. Usually one single Si wafer for a complex circuit presents several levels of lithography with intricate patterns. Nonetheless, the planar process allows several wafers or devices to be fabricated in parallel with high reliability and yield, strongly reducing its cost and production time. Current optical lithography methods [126] are the most widely implemented techniques for exposure, meeting the requirements of large number, low-cost and high reliability, to define features down to approximately 1 μm size. These particularities are still not met by other alternatives methods such as nano imprint, electron-beam and ion-beam lithography, X-ray, nanoindentation or interference lithography methods. The patterning process of a GMR sensor consists of sequential steps of pattern design and transfer as illustrated in Figure 3.17 and detailed described in the following. In summary, the standard procedure to define the sensor element implies a lithographic step in order to imprint the photosensitive polymer (resist) with a certain pattern (mask) and a following step where the pattern is transferred to the GMR thin films.

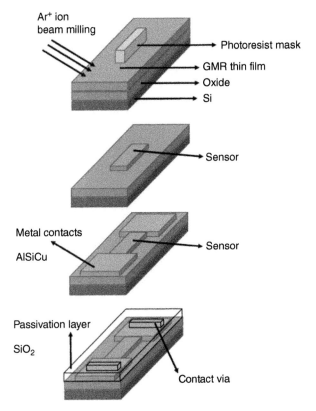

Ar⁺ ion beam milling

Photoresist mask

GMR thin film

Oxide

Si

Sensor

Metal contacts

AlSiCu

Sensor

Passivation layer

SiO₂

Contact via

Figure 3.17 Basic steps for GMR sensor microfabrication: the patterning process of a GMR sensor, sequential steps of pattern design and transfer. *Source:* adapted from [69].

3.5.1.2.1 Photolithography

The photolithography process (Figure 3.18) involves three major steps: (i) coating of the sample with a radiation sensitive polymer solution, called photoresist; (ii) exposure of the resist, patterning a certain design (mask), previously prepared; (iii) development of the transferred pattern.

Coating The resist is spun coated on the surface of the sample, where particular conditions such as coating speed, time, and resist quantity are optimized for the desired thickness of the sensitive layer [127]. The latter is a crucial parameter which influences the lithography resolution [128]. Prior to coating a surface pretreatment is usually required to promote the resist adhesion: a monolayer of hexamethyldisilazane (HMDS) is typically used. After being coated, the resist is soft-baked (typical temperature range of 80–100 °C) to remove solvents and stress while improving adherence.

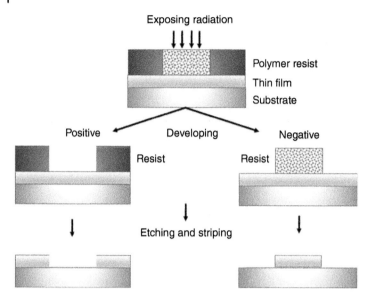

Figure 3.18 Pattern definition by lithography: positive-tone (left figure) or negative-tone (right figure) resist can originate complementary features – holes or pillars. *Source:* adapted from [69].

Lithography The most widely employed standard photolithography systems use a focused laser beam (direct writing systems) or lamps (hard mask aligners) of UV radiation with wavelength typically ranging from 0.5 to 0.1 μm and resolutions below 1 μm are obtained this way [129]. Using hard masks, lithography can be done with the mask as close as possible to the sample (contact lithography) or through an optical system (projection lithography). In this case, hard masks (usually made out of Cr films patterned on quartz) are used with predesigned pattern, which is then transferred to the sample. In direct writing systems the mask is previously elaborated with the aid of CAD tools and then transferred to the wafer using a collimated beam (usually an UV laser). The information from the pattern is in this case codified to an $X-Y$ displacement system, together with an optical turn on/off mechanism. The spot of the light beam moves through the surface in those zones that need to be illuminated, so directly drawing the pattern. This is a much slower system (full exposure of 150-mm wafer area can take 8–16 hours), but still provides a relatively low-cost way for developing prototypes at low-scale production, as no hard masks are needed. On the other hand, if higher resolutions (<500 nm) are required, X-ray, electron, or ion beams systems can be used to transfer the pattern to the resist layer. Electron beam lithography is also particularly used for hard mask design. A good overview of the lithography processes can be found in [130].

Development   After lithography, the exposed patterned is developed: first the sample undergoes a post-bake (typically 80–110 °C) to stop uncompleted resist reactions and remove stress. Then resist developer is sprayed or poured over the substrate. If the resist is positive, exposed regions are rendered soluble and will therefore be removed during development (see Figure 3.18). In contrast, if the resist is negative, exposed regions will harden and remain after developing. In either case, upon developing, the sample is washed to stop the development process and dried, and the pattern is finally printed into the resist layer.

3.5.1.2.2   Pattern Transfer Techniques

After designing the pattern into the resist, one has to transfer it into the underlying film, using either additive techniques such as lift-off, or subtractive methods as etching. Figure 3.19 compares side-by-side the steps involved for both methods.

Patterning Using Lift-off   Being an additive step, the liftoff process starts by defining the resist mask on a substrate and only then depositing the thin film on top (Figure 3.19, right). Afterwards, the sample is placed in a resist stripping solution that will remove the resist layer and all the material placed on top of it,

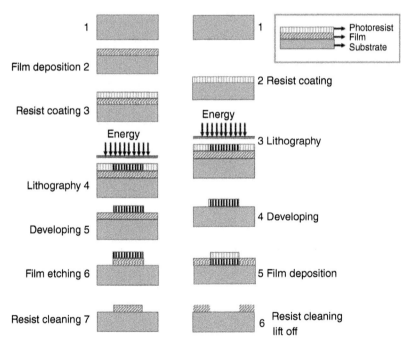

Figure 3.19 Basic steps in two typical pattern transfer techniques: (Left) Using etching for film removing and (right) using liftoff. *Source:* adapted from [69].

leaving the patterned material in the areas previously unoccupied by the mask. This step is widely used for electrical contacts metallization. Lift-off patterning has the advantage of being independent of the material underneath the photo-resist (contrary to etching, where over etch and/or surface oxidation and corrosion can occur), so preventing the substrate from unexpected corrosion. However, it is not possible to lift-off thick films (usually, films with more than half the thickness of the resist layer), nor films deposited with good step coverage (e.g. deposition by CVD).

Etching Etching concerns a process capable of selective removal of undesired portions of a deposited layer. The selective characteristic is provided by the patterned resist mask and also by the properties of the involved layers. The starting point is usually the film to be patterned deposited on a substrate with the desired pattern defined in the top resist mask. Physical dry etching is usually achieved by a controlled removal of material using plasma etching (reactive etching or an ion beam system). In particular Ion Beam etching (ion milling) offers slow (typically below 0.2 nm/s), but very controlled and stable etching ratios and is commonly used for the patterning of GMR devices [118]. Ion-milling etching is an anisotropic process with an etch rate depending on the material type. Moreover, the etched feature profile depends on the angle between the beam and sample, which can be used to control the magnetic properties of patterned materials in sensors or magnetic flux guides [131]. Wet etching concerns a process taking advantage of the corrosive properties of some substances, usually acids. Because inorganic materials such as polymer-based resists are resistant to the inorganic acids action, the wet etching can be performed. The used chemical strongly depends on the material of the layer to be etched (chemically selective), being easily found in the literature with detailed specifications of etchers, time/temperature, according to each material. Due to its aggressive nature and highly isotropic nature, wet etching is not very popular for patterning GMR thin film structures with dimensions ranging below 100 μm.

3.6 Biomedical Magnetoresistive Sensing Applications

Magnetic fields (generated and/or measured) are extensively used in biologic sciences including genetics, biotechnology, different fields of medicine (physiology, oncology, etc.), among others [56]. Most of these applications require the measurement of very low magnetic fields (below the nT limit) in small spaces (commonly in the sub-mm range). At the beginning of the century, MR sensors started to be explored as the sensing elements in biochips. A biosensor can be defined as a "compact analytical device or unit incorporating a biological or biologically-derived sensitive element integrated or associated with a physio-chemical transducer" [132].

3.6.1 Detection of Bioanalytes

General bioanalytes (molecules, cells, viruses, bacteria, tissues) are not magnetic. In order to take advantage of GMR for monitoring or detecting bioanalytes, they must be bonded to magnetic elements, usually nanoparticles (NPs) [133, 134], and driven near the sensor by means of microfluidics [135] or guiding magnetic fields [136, 137]. GMR sensors have been successfully applied to the detection of proteins [138], DNA [139], viruses [140], and bacteria [141]. In this way, two approaches can be defined: static and dynamic. An example of the static approach is the analysis of DNA [142]. For DNA detection, single-stranded DNA receptors are first immobilized on the surface of magnetic sensors. Oligonucleotides of unknown sequence are selectively captured by complementary probes. SA-coated magnetic NPs are then introduced and bind to the biotin of the hybridized DNA. Finally, magnetic field disturbances because of the NPs are sensed by magnetic sensors. Biotin and streptavidin are often used in this process [139]. To improve the performance of the sensor system, micro-coils can be integrated in association with the sensing elements. These coils generate a magnetic field that is used to attract the magnetic beads to the sensor area and activate them [143]. In this way, the femtomolar limit of detection (LOD) has been achieved. For the detection of general cells (cytometry), magnetic nanobeads need to be bonded to them [144]. Then, by means of microfluidics or guiding magnetic fields, they are driven close to the sensors, where the detection is performed, as described in [145].

3.6.2 Monitoring of Magnetic Fluids

Fluids incorporating magnetic particles (usually nanobeads) are known as magnetic fluids. They can be made biocompatible for in vivo applications, such as hyperthermia cancer therapy. A magnetic fluid is injected into the affected area and an external AC magnetic flux density is applied to exploit the self-heating properties of the magnetic beads in the fluid. Temperatures in excess of 42 °C destroy tumors [69]. Hence, the accurate estimation of magnetic fluid content density is critical for the success of the treatment. In [146], a GMR-based needle probe 20 mm long and 310 μm wide, comprising four SV sensors was analyzed. The needle probe was successfully tested in tumor-simulating cylindrical agar cavities.

3.6.3 Biomolecular Recognition Experiments

One of the most important fields, which gathered increasing attention due to the extensive improvements in genome sequencing of the last decades, is the detection of DNA and RNA signatures [147]. Lab-on-a-chip (LOC) platforms for nucleic acid-based detection, find interesting applications in the agrifood industry for the quantification of pathogens and, in clinical diagnosis, for the detection of cancer carrier genic mutation. A typical biochip, employed in molecular recognition experiment, is composed by a biological-sensing system (bioreceptor), a transducer and an output system. The bioreceptor is a biological probe typically immobilized on the biochip surface. Through microspotting, an array of several different probes can be achieved and the whole chip is called a microarray: DNA microarrays with thousands of probes, consisting of single strand DNA (ssDNA), can be used for the study of different genes expressions. In a typical biomolecular recognition detection experiment after the immobilization of the probes, the target sample under inspection is put in contact with the chip surface. This target sample is typically a genomic DNA whose specific genes have been amplified using an amplification technique. After hybridization, a washing step is performed in order to remove all the nonhybridized DNA strands. The stringency control of this washing step is crucial in order to avoid false negative and positive results. Finally, the transducer converts the molecular recognition event into an external signal (optical, electrical, magnetic, etc.). This signal is detected and generally converted into an electrical signal in order to obtain the conclusive data from the experiments.

The detection of the molecular recognition events can be direct (label-free), or by means of additional markers bound to the target molecules. The latter generally have the advantage of large signal-to-noise ratio, since markers generally possess properties which can be easily distinguished from the background. However, the use of markers requires an additional molecular recognition step, which can reduce the overall detection efficiency. MR biosensor arrays combined with magnetic markers have emerged as a new promising platform for biosensing. Following the first demonstration in 1998 [148] driven by the development of GMR-based magnetic field sensors in data storage, MR biosensing was extensively investigated in the past decade, and devices based on AMR [10–12, 149], GMR [150–156], and TMR were developed [53, 157], reaching LOD in the femtomolar [158] and zeptomolar [150] range employing different detection methods and magnetic markers. Among the large family of MR sensors, GMR sensors based on spin-valves and TMR sensors based on MTJs provide the best results in terms of sensitivity and LOD. Other than their extremely high sensitivity and LOD, and fast performance, MR sensors have some advantages, which make them ideal candidates for integration in LOC devices:

- The direct translation of magnetic signal into an electrical one through magnetoresistance makes them readily integrable with conventional electronic platforms.
- Their fabrication which is based on well-established semiconductor processing techniques, allowing for low-cost mass production and high scalability.

- The use of magnetic markers, particularly advantageous because usually all other components in the sample solution are nonmagnetic, thus eliminating interference effects and minimizing the background signal.
- The stability of their magnetic properties over time (in contrast with fluorescent labels which are affected by bleaching and quenching effects).
- The ease in functionalization of their surface with suitable receptors, simplifying specific binding to desired biomolecules.
- The possibility to apply local forces on the markers by generating magnetic gradient fields or exploiting the controlled domain wall motion in magnetic nanostructures, [12, 151, 159, 160] that opens up the interesting option of manipulating molecules through the motion of their labels.

The most widely used scheme, very similar to the fluorescence-based microarrays, relies on the detection of molecular recognition events between the probe molecules, bound onto the sensor surface, and the target molecules which bind specifically to the complementary probes. A washing step then removes the nonhybridized molecules. The magnetic labeling can be performed before (pre-hybridization) or after (post-hybridization) the hybridization step (Figure 3.20). The stray field of the superparamagnetic labels on the surface of the sensor causes a change in the electrical resistance of the sensor which is related to the concentration of the immobilized target molecules. Detection of natural Listeria and the preliminary result in the recognition of genic mutations of KRAS have been demonstrated, using a highly sensitive bio-sensing platform based on MR biosensors integrated in a compact Lab-on-a-chip device [147]. The sketch of a DNA–DNA hybridization event with MR sensors in combination with magnetic marker has been shown in Figure 3.21.

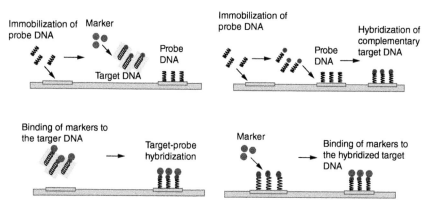

Figure 3.20 A schematic of the pre-hybridization (left) and post-hybridization (right) processes for the not label free method of detection. *Source:* adapted from [147].

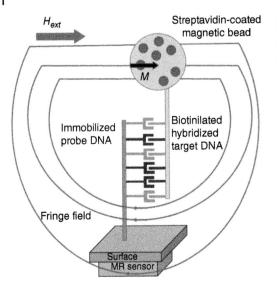

H_{ext}

Streptavidin-coated magnetic bead

M

Immobilized probe DNA

Biotinilated hybridized target DNA

Fringe field

Surface MR sensor

Figure 3.21 Sketch of a DNA–DNA hybridization event with MR sensors in combination with magnetic marker [147].

3.6.4 Ultrasensitive Magnetic Array for Recording of Neuronal Activity (UMANA)

Another area of application of MR sensors in the biological field is neuronal activity sensing where, field signals are of very low intensity and frequency (nT–pT range below1 kHz). In this framework, the knowledge of the brain functionalities from single cell level up to large neuronal ensembles is needed. This implies cross-connecting investigations at the cellular and circuiting level requiring sensing devices with challenging detectivity limits that allow parallel measurements with low invasiveness. Some major breakthroughs have already been achieved, leading the way to portable neural activity sensing [161–163]. The main current method for recording and stimulating the electric activity of single cells is patch-clamp, in which a borosilicate micropipette (containing the electrode) is tightly sealed to the cell membrane [164]. For physiological and pharmacological purposes, patch-clamp is very sensitive; however, it presents many drawbacks. It requires considerable micromanipulation skills (particularly to record in brain slices), it does not allow to record simultaneously from more than a couple of neurons and finally it is not free of artefacts (e.g. because of exchange between the pipette solution and the cytoplasm). Alternative methods have been recently developed to record from neuronal networks, such as extracellular multielectrode arrays (MEA). Neuronal cultures or acute slices are laid onto the electrode-containing plates and independent recording (or stimulation) can be carried out from up to 256 electrodes. In this case, because a single electrode samples from a field that usually arises from several cells, it may be difficult to establish a one-to-one correlation between the recorded trace and a specific neuron [165]. More recently, various high-density (HD) MEAs with higher resolution by exploiting Complementary Metal Oxide Semiconductor (CMOS) technologies have been developed and validated [166, 167]. However, also in this case it is generally difficult to establish a one-to-one correlation between the recorded trace and a specific neuron, because a single electrode pick-up an electric signal usually arising from several cells. Other emergent techniques present also relevant drawbacks. Voltage-sensitive dyes used to monitor membrane potential have poor signal-to-noise ratio and are toxic, thus preventing long-term recordings [168]. Voltage-sensitive fluorescent proteins, on the other hand, present limited spatial and temporal resolution [166, 169]. An alternative strategy is generating nanotechnology-based on-chip platforms for the investigation of neuronal activity, which is a major current topic in neuroscience. However, integrating nanotechnology and biology is anything but easy. Sensitive transducers to measure the neuronal signals are necessary together with methods to properly position the neurons on the chip surface and produce an appropriate microenvironment to ensure neuronal survival and efficient coupling with the sensors. Nano-sized field effect transistors [170], nanopillars [171], and extracellular techniques demonstrated good temporal resolution and signal-to-noise ratio, but their spatial resolution is inherently limited by the slow spatial decay of electric fields generated by the membrane potentials.

Recently, novel methods have been proposed for magnetic recording of the neuronal activity: from brain slices using MR sensors [161] and, at the subcellular level, using quantum Nitrogen Vacancy in diamond (NV-diamond) [172]. The main advantages of magnetic detection are the noninvasiveness and the high spatial resolution due to the inherent localization of magnetic fields generated by neuronal currents, thus allowing to record not only from single neurons but also from subcellular compartments (axon and axonal initial segment, synaptic contacts, and neuronal soma). A possible example of the chip layout is a neuronal cell grown properly aligned to a dense Multi Magnetic sensor Array (MMA), which in this case is able to record the magnetic signal arising from the ionic currents pulses along the axon. Such a block will be replicated on the same chip to maximize the throughput in case of experiments on single neural cells or networks, combining single cell sensitivity and multiplexing. Highly sensitive magnetic sensors based on MTJs or GMR can be employed to detect the local magnetic fields generated by the neuronal currents.

Baselt et al. described a new concept in biological labeling and magnetic sensor detection based on GMR sensors [148]. They introduced a semiconductor-based multilayer GMR sensor chip, which came to be known as the bead array counter chip that detects local in-plane magnetic fields produced by paramagnetic microbeads immobilized directly above the sensor surface during binding assays. Other investigators have also followed Baselt's initial approach with other GMR sensor devices [154, 173]. GMR spin valve device has been used for magnetically labeled biosensors by Graham et al. [132, 159]. Graham et al. described a 2 µm × 6 µm sensor consisting of a MR material stack with two FM layers, typically a NiFe-based composite, which are separated by a Cu spacer. Recent research announced that a graphene layer can be used as the separation layer for a GMR spin-valve [174]. However, the spin-valve with a middle layer of graphene is more like a TMR, because the graphene is not conductive in its perpendicular direction.

References

1 Freitas, P.P., Ferreira, R., Cardoso, S., and Cardoso, F. (2007). Magnetoresistive sensors. *Journal of Physics: Condensed Matter* 19 (16): 165221.

2 Thomson, W. (1857). On the electro-dynamic qualities of metals: effects of magnetization on the electric conductivity of nickel and iron. *Proceedings of the Royal Society of London* 8: 546–550.

3 Mcguire, T.R. and Potter, R.I. (1975). Anisotropic magnetoresistance in ferromagnetic 3d alloys. *IEEE Transactions on Magnetics* 11 (4): 1018–1038.

4 Beguivin, A., Corte-León, H., Manzin, A. et al. (2014). Simultaneous magnetoresistance and magneto-optical measurements of domain wall properties in nanodevices. *Journal of Applied Physics* 115 (17): 17C718.

5 Corte-León, H., Nabaei, V., Manzin, A. et al. (2014). Anisotropic magnetoresistance state space of permalloy nanowires with domain wall pinning geometry. *Scientific Reports* 4: 6045.

6 Jan, J.P. (1957). Galvamomagnetic and thermomagnetic effects in metals. In: *Solid State Physics*, vol. 5 (eds. F. Seitz and D. Turnbull), 1–96. Elsevier.

7 Vopálenský, M. (2014). Measuring with magnetoresistive sensors. PhD thesis, Faculty of Electrical Engineering, Czech Technical University in Prague, Czech Republic. https://dspace.cvut.cz/bitstream/handle/10467/52742/Habilitace_Vopalensky_2014.pdf?sequence=1&isAllowed=y (accessed 2 July 2019).

8 Tumanski, S. (2001). *Thin Film Magnetoresistive Sensors*. CRC Press.

9 Jander, A., Smith, C., and Schneider, R. (2005). Magnetoresistive sensors for nondestructive evaluation. *Proceedings of the 10th SPIE International Symposium on Nondestructive Evaluation for Health Monitoring and Diagnostics*, San Diego, California, USA.

10 Miller, M.M., Prinz, G.A., Cheng, S.F., and Bounnak, S. (2002). Detection of a micron-sized magnetic sphere using a ring-shaped anisotropic magnetoresistance-based sensor: a model for a magnetoresistance-based biosensor. *Applied Physics Letters* 81 (12): 2211–2213.

11 Donolato, M., Gobbi, M., Vavassori, P. et al. (2009). Nanosized corners for trapping and detecting magnetic nanoparticles. *Nanotechnology* 20 (38): 385501.

12 Vavassori, P., Metlushko, V., Ilic, B. et al. (2008). Domain wall displacement in Py square ring for single nanometric magnetic bead detection. *Applied Physics Letters* 93 (20): 203502.

13 Vieira, G., Chen, A., Henighan, T. et al. (2012). Transport of magnetic microparticles via tunable stationary magnetic traps in patterned wires. *Physical Review B* 85 (17): 174440.

14 Donolato, M., Vavassori, P., Gobbi, M. et al. (2010). On-chip manipulation of protein-coated magnetic beads via domain-wall conduits. *Advanced Materials* 22 (24): 2706.

15 Ruan, G., Vieira, G., Henighan, T. et al. (2010). Simultaneous magnetic manipulation and fluorescent tracking of multiple individual hybrid nanostructures. *Nano Letters* 10 (6): 2220–2224.

16 Vieira, G., Henighan, T., Chen, A. et al. (2009). Magnetic wire traps and programmable manipulation of biological cells. *Physical Review Letters* 103 (12): 128101.

17 Baibich, M.N., Broto, J.M., Fert, A. et al. (1988). Giant magnetoresistance of (001)Fe/(001)Cr magnetic superlattices. *Physical Review Letters* 61 (21): 2472–2475.

18 Binasch, G., Grunberg, P., Saurenbach, F., and Zinn, W. (1989). Enhanced magnetoresistance in layered magnetic-structures with antiferromagnetic interlayer exchange. *Physical Review B* 39 (7): 4828–4830.

19 White, R.L. (1992). Giant magnetoresistance: a primer. *IEEE Transactions on Magnetics* 28 (5): 2482–2487.

20 Rizal, C. (2012). *Giant Magnetoresistance and Magnetic Properties of Ferromagnetic Hybrid Nanostructures*. University of British Columbia.

21 Inoue, J.-I., Oguri, A., and Maekawa, S. (1991). Theory of giant magnetoresistance in metallic superlattices. *Journal of the Physical Society of Japan* 60 (2): 376–379.

22 Bakonyi, I. and Péter, L. (2010). Electrodeposited multilayer films with giant magnetoresistance (GMR): progress and problems. *Progress in Materials Science* 55 (3): 107–245.

23 Ferrari, E.F., Da Silva, F.C.S., and Knobel, M. (1999). Theory of giant magnetoresistance in granular alloys. *Physical Review B* 59 (13): 8412.

24 Berkowitz, A.E., Mitchell, J.R., Carey, M.J. et al. (1992). Giant magnetoresistance in heterogeneous Cu–Co alloys. *Physical Review Letters* 68 (25): 3745.

25 Berkowitz, A.E., Mitchell, J.R., Carey, M.J. et al. (1993). Giant magnetoresistance in heterogeneous Cu–Co and Ag–Co alloy films. *Journal of Applied Physics* 73 (10): 5320–5325.

26 Rizal, C. (2010). Magnetoresistance and magnetic properties of Co (tCo)/Cu multilayer films. *IEEE Transactions on Magnetics* 46 (6): 2394–2396.

27 Rizal, C., Ueda, Y., and Pokharel, R.K. (2011). Magnetotransport properties of Co–Au granular alloys. *International Journal of Applied Physics and Mathematics* 1 (3): 161.

28 Kline, T.L., Xu, Y.-H., Jing, Y., and Wang, J.-P. (2009). Biocompatible high-moment FeCo–Au magnetic nanoparticles for magnetic hyperthermia treatment optimization. *Journal of Magnetism and Magnetic Materials* 321 (10): 1525–1528.

29 Zaman, H., Ikeda, S., and Ueda, Y. (1997). Magnetoresistance in Co–Ag multilayers and granular films produced by electrodeposition method. *IEEE Transactions on Magnetics* 33 (5): 3517–3519.

30 Ueda, Y., Hataya, N., and Zaman, H. (1996). Magnetoresistance effect of Co/Cu multilayer film produced by electrodeposition method. *Journal of Magnetism and Magnetic Materials* 156 (1–3): 350–352.

31 Ueda, Y., Ikeda, S., Mori, Y., and Zaman, H. (1996). Magnetoresistance and magnetism in Fe–Cu alloys produced by electrodeposition and mechanical alloying methods. *Materials Science and Engineering: A* 217: 371–375.

32 Wang, J.-P., Shen, W., and Bai, J. (2005). Exchange coupled composite media for perpendicular magnetic recording. *IEEE Transactions on Magnetics* 41 (10): 3181–3186.

33 Shen, W.K., Das, A., Racine, M. et al. (2006). Enhancement in magnetic anisotropy for hcp-structured Co alloy thin films through Pt addition. *IEEE Transactions on Magnetics* 42 (10): 2945–2947.

34 Ueda, Y., Houga, T., Zaman, H., and Yamada, A. (1999). Magnetoresistance effect of Co–Cu nanostructure prepared by electrodeposition method. *Journal of Solid State Chemistry* 147 (1): 274–280.

35 Zaman, H., Yamada, A., Fukuda, H., and Ueda, Y. (1998). Magnetoresistance effect in Co–Ag and Co–Cu alloy films prepared by electrodeposition. *Journal of the Electrochemical Society* 145 (2): 565–568.

36 Xiao, J.Q., Jiang, J.S., and Chien, C.L. (1992). Giant magnetoresistance in the granular Co–Ag system. *Physical Review B* 46 (14): 9266.

37 Takanashi, K., Park, J., Sugawara, T. et al. (1996). Giant magnetoresistance and microstructure in Cr–Fe and Cu–Co heterogeneous alloys. *Thin Solid Films* 275 (1–2): 106–110.

38 Ueda, Y. and Ito, M. (1994). Magnetoresistance in Co–Cu alloy films formed by electrodeposition method. *Japanese Journal of Applied Physics* 33 (10A): L1403.

39 Kataoka, N., Takeda, H., Echigoya, J. et al. (1995). GMR and micro-structure in bulk Au·Co nanogranular alloys. *Journal of Magnetism and Magnetic Materials* 140: 621–622.

40 Ikeda, S., Houga, T., Takakura, W., and Ueda, Y. (1996). Magnetoresistance in $(Co_xFe_{1-x})_{20}Cu_{80}$ granular alloys produced by mechanical alloying. *Materials Science and Engineering: A* 217: 376–380.

41 Guan, M. and Podlaha, E.J. (2007). Electrodeposition of AuCo alloys and multilayers. *Journal of Applied Electrochemistry* 37 (5): 549–555.

42 Tedrow, P.M. and Meservey, R. (1971). Spin-dependent tunneling into ferromagnetic nickel. *Physical Review Letters* 26 (4): 192.

43 Julliere, M. (1975). Tunneling between ferromagnetic films. *Physics Letters A* 54 (3): 225–226.

44 Moodera, J.S., Kinder, L.R., Wong, T.M., and Meservey, R. (1995). Large magnetoresistance at room-temperature in ferromagnetic thin-film tunnel-junctions. *Physical Review Letters* 74 (16): 3273–3276.

45 Miyazaki, T. and Tezuka, N. (1995). Giant magnetic tunneling effect in Fe/Al_2O_3/Fe junction. *Journal of Magnetism and Magnetic Materials* 139 (3): L231–L234.

46 Chappert, C., Fert, A., and Van Dau, F.N. (2007). The emergence of spin electronics in data storage. *Nature Materials* 6 (11): 813.

47 Li, Z.S., de Groot, C., and Moodera, J.H. (2000). Gallium oxide as an insulating barrier for spin-dependent tunneling junctions. *Applied Physics Letters* 77 (22): 3630–3632.

48 Khan, M.N., Henk, J., and Bruno, P. (2008). Anisotropic magnetoresistance in Fe/MgO/Fe tunnel junctions. *Journal of Physics-Condensed Matter* 20 (15).

49 Zuo, S., Nazarpour, K., and Heidari, H. (2018). Device modeling of MgO-barrier tunneling magnetoresistors for hybrid spintronic-CMOS. *IEEE Electron Device Letters* 39 (11): 1784–1787.

50 Cobas, E., Friedman, A.L., van't Erve, O.M.J. et al. (2012). Graphene as a tunnel barrier: graphene-based magnetic tunnel junctions. *Nano Letters* 12 (6): 3000–3004.

51 Parkin, S.S.P., Kaiser, C., Panchula, A. et al. (2004). Giant tunnelling magnetoresistance at room temperature with MgO (100) tunnel barriers. *Nature Materials* 3 (12): 862–867.

52 Brzeska, M., Panhorst, M., Kamp, P.B. et al. (2004). Detection and manipulation of biomolecules by magnetic carriers. *Journal of Biotechnology* 112 (1–2): 25–33.

53 Shen, W.F., Liu, X.Y., Mazumdar, D., and Xiao, G. (2005). In situ detection of single micron-sized magnetic beads using magnetic tunnel junction sensors. *Applied Physics Letters* 86 (25): 253901.

54 Wang, S.X., Bae, S.Y., Li, G.X. et al. (2005). Towards a magnetic microarray for sensitive diagnostics. *Journal of Magnetism and Magnetic Materials* 293 (1): 731–736.

55 Swagten, H.M. and Paluskar, P.V. (2010). Magnetic tunnel junctions. In: *Encyclopedia of Materials: Science & Technology* (ed. K.H.J. Buschow, R. Cahn, M. Flemings, et al.), 1–7. Elsevier.

56 Francis, L.A. and Poletkin, K. (2017). *Magnetic Sensors and Devices: Technologies and Applications.* CRC Press.

57 Hertz, J.A. and Aoi, K. (1973). Spin-dependent tunnelling from transition-metal ferromagnets. *Physical Review B* 8 (7): 3252.

58 Stearns, M.B. (1977). Simple explanation of tunneling spin-polarization of Fe, Co, Ni and its alloys. *Journal of Magnetism and Magnetic Materials* 5 (2): 167–171.

59 Belashchenko, K.D., Tsymbal, E.Y., Oleynik, I.I., and van Schilfgaarde, M. (2005). Positive spin polarization in Co/Al$_2$O$_3$/Co tunnel junctions driven by oxygen adsorption. *Physical Review B* 71 (22): 224422.

60 Yuasa, S., Sato, T., Tamura, E. et al. (2000). Magnetic tunnel junctions with single-crystal electrodes: a crystal anisotropy of tunnel magneto-resistance. *EPL (Europhysics Letters)* 52 (3): 344.

61 LeClair, P., Swagten, H.J.M., Kohlhepp, J.T. et al. (2000). Apparent spin polarization decay in Cu-dusted Co/Al$_2$O$_3$/Co tunnel junctions. *Physical Review Letters* 84 (13): 2933.

62 LeClair, P., Kohlhepp, J.T., Swagten, H.J.M., and de Jonge, W.J.M. (2001). Interfacial density of states in magnetic tunnel junctions. *Physical Review Letters* 86 (6): 1066.

63 Yuasa, S., Nagahama, T., and Suzuki, Y. (2002). Spin-polarized resonant tunneling in magnetic tunnel junctions. *Science* 297 (5579): 234–237.

64 Kaiser, C., Panchula, A.F., and Parkin, S.S.P. (2005). Finite tunneling spin polarization at the compensation point of rare-earth-metal–transition-metal alloys. *Physical Review Letters* 95 (4): 047202.

65 Kaiser, C., van Dijken, S., Yang, S.-H. et al. (2005). Role of tunneling matrix elements in determining the magnitude of the tunneling spin polarization of 3d transition metal ferromagnetic alloys. *Physical Review Letters* 94 (24): 247203.

66 Sharma, M., Wang, S.X., and Nickel, J.H. (1999). Inversion of spin polarization and tunneling magnetoresistance in spin-dependent tunneling junctions. *Physical Review Letters* 82 (3): 616.

67 De Teresa, J.M., Barthélémy, A., Fert, A. et al. (1999). Role of metal-oxide interface in determining the spin polarization of magnetic tunnel junctions. *Science* 286 (5439): 507–509.

68 Velev, J.P., Belashchenko, K.D., Stewart, D.A. et al. (2005). Negative spin polarization and large tunneling magnetoresistance in epitaxial Co|SrTiO$_3$|Co magnetic tunnel junctions. *Physical Review Letters* 95 (21): 216601.

69 Reig, C., Cardoso, S., and Mukhopadhyay, S.C. (2013). Giant magnetoresistance (GMR) sensors. *Ssmi6*: 157–180.

70 Jeng, J.-T., Chiang, C.-Y., Chang, C.-H., and Lu, C.-C. (2014). Vector magnetometer with dual-bridge GMR sensors. *IEEE Transactions on Magnetics* 50 (1): 1–4.

71 Luong, V., Jeng, J., Lai, B. et al. (2015). D|esign of three-dimensional magnetic field sensor with single bridge of spin-valve giant magnetoresistance films. In: *2015 IEEE International Magnetics Conference (INTERMAG)*, 1–1. IEEE.

72 Chiang, C.-Y., Jeng, J.-T., Lai, B.-L. et al. (2015). Tri-axis magnetometer with in-plane giant magnetoresistance sensors for compass application. *Journal of Applied Physics* 117 (17): 17A321.

73 Mao, S., Gangopadhyay, S., Amin, N., and Murdock, E. (1996). NiMn-pinned spin valves with high pinning field made by ion beam sputtering. *Applied Physics Letters* 69 (23): 3593–3595.

74 Saito, M., Hasegawa, N., Koike, F. et al. (1999). PtMn single and dual spin valves with synthetic ferrimagnet pinned layers. *Journal of Applied Physics* 85 (8): 4928–4930.

75 Lenssen, K.M., Adelerhof, D.J., Gassen, H.J. et al. (2000). Robust giant magnetoresistance sensors. *Sensors and Actuators A: Physical* 85 (1–3): 1–8.

76 Giebeler, C., Adelerhof, D.J., Kuiper, A.E.T. et al. (2001). Robust GMR sensors for angle detection and rotation speed sensing. *Sensors and Actuators A: Physical* 91 (1–2): 16–20.

77 Prakash, S., Pentek, K., and Zhang, Y. (2001). Reliability of PtMn-based spin valves. *IEEE Transactions on Magnetics* 37 (3): 1123–1131.

78 Wienecke, A., Wurz, M.C., and Rissing, L. (2013). Integrierte Sensorik für Hochtemperaturumgebungen [integrated sensor systems for high temperature environments]. GMM, VDI/VDE-IT (eds.), *Tagungsband: Mikrosystemtechnik-Kongress. Von Bauelementen zu Systemen*, Aachen, Deutschland.

79 Wienecke, A. and Rissing, L. (2015). Relationship between thermal stability and layer-stack/structure of NiMn-based GMR systems. *IEEE Transactions on Magnetics* 51 (1): 1–4.

80 Slatter, R. (2015). Magnetoresistive (MR) sensors for angle-, path- and current measurement in harsh environments. *Proceedings of the 17th International Conference on Sensors and Measurement Technology*, Nürnberg, Germany (19–21 May 2015).

81 Schäfer, I. (2011). Aviation sensor requirements: do they fit to MR technology? *Proceedings of 11th MR Symposium*, Wetzlar, Germany.

82 Reig, C., Cubells-Beltrán, M.-D., and Ramírez Muñoz, D. (2009). Magnetic field sensors based on giant magnetoresistance (GMR) technology: applications in electrical current sensing. *Sensors* 9 (10): 7919–7942.

83 Valenzuela, R., Freijo, J.J., Salcedo, A. et al. (1997). A miniature DC current sensor based on magnetoimpedance. *Journal of Applied Physics* 81 (8): 4301–4303.

84 Sebastia, J.P., Muñoz, D.R., de Freitas, P.J.P., and Ku, W. (2004). A novel spin-valve bridge sensor for current sensing. *IEEE Transactions on Instrumentation and Measurement* 53 (3): 877–880.

85 Reig, C., Ramırez, D., Silva, F. et al. (2004). Design, fabrication, and analysis of a spin-valve based current sensor. *Sensors and Actuators A: Physical* 115 (2-3): 259–266.

86 Pelegrí-Sebastiá, J. and Ramírez-Muñoz, D. (2003). Safety device uses GMR sensor. *Edn* 48 (15): 84–84.

87 Pelegrí, J., Ejea, J.B., Ramirez, D., and Freitas, P.P. (2003). Spin-valve current sensor for industrial applications. *Sensors and Actuators A: Physical* 105 (2): 132–136.

88 Muñoz, D.R., Pérez, D.M., Moreno, J.S. et al. (2009). Design and experimental verification of a smart sensor to measure the energy and power consumption in a one-phase AC line. *Measurement* 42 (3): 412–419.

89 Vopálenský, M., Platil, A., and Kašpar, P. (2005). Wattmeter with AMR sensor. *Sensors and Actuators A: Physical* 123: 303–307.

90 Ramírez, D. and Pelegrí, J. (1999). GMR sensors manage batteries. *Edn* 44 (138): 31.

91 Reig, C., Ramírez, D., Li, H.H., and Freitas, P.P. (2005). Low-current sensing with specular spin valve structures. *IEE Proceedings-Circuits, Devices and Systems* 152 (4): 307–311.

92 Muñoz, D.R., Berga, S.C., and Escrivá, C.R. (2005). Current loop generated from a generalized impedance converter: a new sensor signal conditioning circuit. *Review of Scientific Instruments* 76 (6): 066103.

93 Cubells-Beltrán, M.D., Reig, C., Muñoz, D.R. et al. (2009). Full Wheatstone bridge spin-valve based sensors for IC currents monitoring. *IEEE Sensors Journal* 9 (12): 1756–1762.

94 Granig, W., Weinberger, M., Reidl, C. et al. (2010). Integrated GMR angle sensor for electrical commutated motors including features for safety critical applications. *Procedia Engineering* 5: 1384–1387.

95 Burke, J., Moynihan, J.F., and Unterkofler, K. (2000). Extraction of high resolution position information from sinusoidal encoders. In: *Proceedings of the International Intelligent Motion Conference*, 217–222. Intertec International, Inc.

96 Jogschies, L., Klaas, D., Kruppe, R. et al. (2015). Recent developments of magnetoresistive sensors for industrial applications. *Sensors* 15 (11): 28665–28689.

97 Overmeyer, L., Rissing, L., Wurz, M.C. et al. (eds.) (2011). Component-integrated sensors and communication for gentelligent devices. In: *2011 IEEE International Conference on Industrial Engineering and Engineering Management (IEEM)*, 499–503. IEEE.

98 Wu, K.H. and Gatzen, H.H. (2008). Development of a system for data storage on machine components. *Proceedings of the I PROMS 2008, the 4th Virtual Conference of the EU-Funded FP6 I PROMS Network of Excellence on Innovative Production Machines and Systems*, Wales, UK (1–14 July 2008), pp. 454–459.

99 Taptimthong, P., Rittinger, J., Wurz, M.C., and Rissing, L. (2014). Flexible magnetic writing/reading system: polyimide film as flexible substrate. *Procedia Technology* 15: 230–237.

100 Nabaei, V., Chandrawati, R., and Heidari, H. (2018). Magnetic biosensors: modelling and simulation. *Biosensors and Bioelectronics* 103: 69–86.

101 Li, F. and Kosel, J. (2013). A magnetic biosensor system for detection of *E. coli*. *IEEE Transactions on Magnetics* 49 (7): 3492–3495.

102 Li, F. and Kosel, J. (2014). An efficient biosensor made of an electromagnetic trap and a magneto-resistive sensor. *Biosensors and Bioelectronics* 59: 145–150.

103 Li, F. and Kosel, J. (2012). A magnetic method to concentrate and trap biological targets. *IEEE Transactions on Magnetics* 48 (11): 2854–2856.

104 Liu, Y., Jin, W., Yang, Y., and Wang, Z. (2006). Micromagnetic simulation for detection of a single magnetic microbead or nanobead by spin-valve sensors. *Journal of Applied Physics* 99 (8): 08G102.

105 Li, G., Joshi, V., White, R.L. et al. (2003). Detection of single micron-sized magnetic bead and magnetic nanoparticles using spin valve sensors for biological applications. *Journal of Applied Physics* 93 (10): 7557–7559.

106 Donahue, M., Porter, D., Lau, J. et al. (1999). Interagency report NISTIR 6376. National institute of standards and technology, Gaithersburg. *NIST Journal of Research* 114: 57–67.

107 Williams, E.M. (2001). *Design and Analysis of Magnetoresistive Recording Heads*. Wiley.

108 Li, Y., Wang, Y., Klein, T., and Wang, J.-P. (2014). External-field-free magnetic biosensor. *Applied Physics Letters* 104 (12): 122401.

109 Donahue, M.J. and Porter, D.G. (1999). *OOMMF User's Guide, Version 1.0*. NIST Interagency/Internal Report (NISTIR) 6376. Gaithersburg, MD: NIST.

110 Wirix-Speetjens, R., Fyen, W., De Boeck, J., and Borghs, G. (2006). Single magnetic particle detection: experimental verification of simulated behavior. *Journal of Applied Physics* 99 (10): 103903.

111 O'Handley, R.C. (2000). *Modern Magnetic Materials: Principles and Applications*. Wiley.

112 Hirota, E., Sakakima, H., and Inomata, K. (2013). *Giant Magneto-Resistance Devices*. Springer Science & Business Media.

113 Jaeger, R.C. (1993). *Introduction to Microelectronic Fabrication*, Addison-Wesley Modular Series on Solid State Devices, vol. 5 (eds. G.W. Neudeck and R.F. Pierret). Addison-Wesley Publishing Co., Inc.

114 Cheng, N., Ahn, J., and Krishnan, K.M. (2001). Epitaxial growth and exchange biasing of PdMn/Fe bilayers grown by ion-beam sputtering. *Journal of Applied Physics* 89 (11): 6597–6599.

115 Schwebel, C. and Gautherin, G. (1988). Deposition of thin films by ion beam sputtering: mechanisms and epitaxial growth. *AIP Conference Proceedings* 167: 237–249.

116 Gehanno, V., Freitas, P.P., Veloso, A. et al. (1999). Ion beam deposition of Mn-Ir spin valves. *IEEE Transactions on Magnetics* 35 (5): 4361–4367.

117 Dong, L. and Srolovitz, D.J. (1998). Texture development mechanisms in ion beam assisted deposition. *Journal of Applied Physics* 84 (9): 5261–5269.

118 Nordiko Technical Services. http://www.nordiko-tech.com/products.html (accessed 2 July 2019).

119 Veeco. http://www.veeco.com/products/ion-beam.aspx (accessed 2 July 2019).

120 Ferreira, R., Cardoso, S., Freitas, P.P. et al. (2012). Influence of ion beam assisted deposition parameters on the growth of MgO and CoFeB. *Journal of Applied Physics* 111 (7): 07C117.

121 Cardoso, S., Gehanno, V., Ferreira, R., and Freitas, P.P. (1999). Ion beam deposition and oxidation of spin-dependent tunnel junctions. *IEEE Transactions on Magnetics* 35 (5): 2952–2954.

122 Pierson, H.O. (1999). *Handbook of Chemical Vapor Deposition: Principles, Technology and Applications*. William Andrew.

123 Schlesinger, M. and Paunovic, M. (2000). *Modern Electroplating*, 5e. New York: Wiley.

124 Whitney, T.M., Searson, P.C., Jiang, J.S., and Chien, C.L. (1993). Fabrication and magnetic properties of arrays of metallic nanowires. *Science* 261 (5126): 1316–1319.

125 Evans, P.R., Yi, G., and Schwarzacher, W. (2000). Current perpendicular to plane giant magnetoresistance of multilayered nanowires electrodeposited in anodic aluminum oxide membranes. *Applied Physics Letters* 76 (4): 481–483.

126 Jaeger, R.C. (2001). *Introduction to Microelectronic Fabrication: Volume 5 of Modular Series on Solid State Devices*. Upper Saddle River: Prentice Hall.

127 Middleman, S. and Hochberg, A.K. (1993). *Process Engineering Analysis in Semiconductor Device Fabrication*. New York: McGraw-Hill.

128 Leitao, D.C., Macedo, R.J., Silva, A.V. et al. (eds.) (2012). Optimization of exposure parameters for lift-off process of sub-100 features using a negative tone electron beam resist. In: *2012 12th IEEE International Conference on Nanotechnology (IEEE-NANO)*, 1–6. IEEE.

129 Nabiyouni, G.R. (2009). Design and fabrication of nanomagnetic sensors based on electrodeposited GMR materials. *Metrology and Measurement Systems* XVI (3): 519–529.

130 Waser, R. (2012). *Nanoelectronics and Information Technology*. Wiley.

131 Amaral, J., Gaspar, J., Pinto, V. et al. (2013). Measuring brain activity with magnetoresistive sensors integrated in micromachined probe needles. *Applied Physics A* 111 (2): 407–412.

132 Graham, D.L., Ferreira, H.A., and Freitas, P.P. (2004). Magnetoresistive-based biosensors and biochips. *Trends in Biotechnology* 22 (9): 455–462.

133 Wang, W., Wang, Y., Tu, L. et al. (2014). Magnetoresistive performance and comparison of supermagnetic nanoparticles on giant magnetoresistive sensor-based detection system. *Scientific Reports* 4: 5716.

134 Lee, K., Lee, S., Cho, B.K. et al. (2009). The limit of detection of giant magnetoresistive (GMR) sensors for bio-applications. *Journal of Korean Physical Society* 55: 193.

135 Muluneh, M. and Issadore, D. (2014). A multi-scale PDMS fabrication strategy to bridge the size mismatch between integrated circuits and microfluidics. *Lab on a Chip* 14 (23): 4552–4558.

136 Giouroudi, I. and Keplinger, F. (2013). Microfluidic biosensing systems using magnetic nanoparticles. *International Journal of Molecular Sciences* 14 (9): 18535–18556.

137 Gooneratne, C.P., Liang, C., and Kosel, J. (2011). A planar conducting microstructure to guide and confine magnetic beads to a sensing zone. *Microelectronic Engineering* 88 (8): 1757–1760.

138 Gaster, R.S., Xu, L., Han, S.-J. et al. (2011). Quantification of protein interactions and solution transport using high-density GMR sensor arrays. *Nature Nanotechnology* 6 (5): 314.

139 Wang, W., Wang, Y., Tu, L. et al. (2013). Surface modification for protein and DNA immobilization onto GMR biosensor. *IEEE Transactions on Magnetics* 49 (1): 296–299.

140 Zhi, X., Deng, M., Yang, H. et al. (2014). A novel HBV genotypes detecting system combined with microfluidic chip, loop-mediated isothermal amplification and GMR sensors. *Biosensors and Bioelectronics* 54: 372–377.

141 Mujika, M., Arana, S., Castano, E. et al. (2009). Magnetoresistive immunosensor for the detection of *Escherichia coli* O157: H7 including a microfluidic network. *Biosensors and Bioelectronics* 24 (5): 1253–1258.

142 Koets, M., Van der Wijk, T., Van Eemeren, J. et al. (2009). Rapid DNA multi-analyte immunoassay on a magneto-resistance biosensor. *Biosensors and Bioelectronics* 24 (7): 1893–1898.

143 Freitas, P.P., Cardoso, S., Ferreira, R. et al. (2011). Optimization and integration of magnetoresistive sensors. *Spin* 1: 71–91.

144 Freitas, P.P., Cardoso, F.A., Martins, V.C. et al. (2012). Spintronic platforms for biomedical applications. *Lab on a Chip* 12 (3): 546–557.

145 Shoshi, A., Schotter, J., Schroeder, P. et al. (2012). Magnetoresistive-based real-time cell phagocytosis monitoring. *Biosensors and Bioelectronics* 36 (1): 116–122.

146 Mukhopadhyay, S.C., Chomsuwan, K., Gooneratne, C.P., and Yamada, S. (2007). A novel needle-type SV-GMR sensor for biomedical applications. *IEEE Sensors Journal* 7 (3): 401–408.

147 Scolari, M. (2015). Magnetoresistive sensors for biological applications. PhD thesis, Politecnico di Milano.

148 Baselt, D.R., Lee, G.U., Natesan, M. et al. (1998). A biosensor based on magnetoresistance technology. *Biosensors & Bioelectronics* 13 (7–8): 731–739.

149 Ejsing, L., Hansen, M.F., Menon, A.K. et al. (2004). Planar Hall effect sensor for magnetic micro-and nanobead detection. *Applied Physics Letters* 84 (23): 4729–4731.

150 Gaster, R.S., Hall, D.A., and Wang, S.X. (2011). nanoLAB: an ultraportable, handheld diagnostic laboratory for global health. *Lab on a Chip* 11 (5): 950–956.

151 Ferreira, H.A., Feliciano, N., Graham, D.L. et al. (2005). Rapid DNA hybridization based on AC field focusing of magnetically labeled target DNA. *Applied Physics Letters* 87 (1): 013901.

152 Graham, D.L., Ferreira, H., Bernardo, J. et al. (2002). Single magnetic microsphere placement and detection on-chip using current line designs with integrated spin valve sensors: biotechnological applications. *Journal of Applied Physics* 91 (10): 7786–7788.

153 Miller, M.M., Sheehan, P.E., Edelstein, R.L. et al. (2001). A DNA array sensor utilizing magnetic microbeads and magnetoelectronic detection. *Journal of Magnetism and Magnetic Materials* 225 (1–2): 138–144.

154 Schotter, J., Kamp, P.B., Becker, A. et al. (2002). A biochip based on magnetoresistive sensors. *IEEE Transactions on Magnetics* 38 (5): 3365–3367.

155 Yin, Z., Bonizzoni, E., and Heidari, H. (2018). Magnetoresistive biosensors for on-chip detection and localization of paramagnetic particles. *IEEE Journal of Electromagnetics, RF and Microwaves in Medicine and Biology* 2 (3): 179–185.

156 Samyak, S. and Heidari, H. (2017). On-chip magnetoresistive sensors for detection and localization of paramagnetic particles. *Proceedings of IEEE SENSORS Conference*, Glasgow, UK.

157 Shen, W., Schrag, B.D., Carter, M.J., and Xiao, G. (2008). Quantitative detection of DNA labeled with magnetic nanoparticles using arrays of MgO-based magnetic tunnel junction sensors. *Applied Physics Letters* 93 (3): 033903.

158 Martins, V.C., Cardoso, F.A., Germano, J. et al. (2009). Femtomolar limit of detection with a magnetoresistive biochip. *Biosensors and Bioelectronics* 24 (8): 2690–2695.

159 Graham, D.L., Ferreira, H.A., Feliciano, N. et al. (2005). Magnetic field-assisted DNA hybridisation and simultaneous detection using micron-sized spin-valve sensors and magnetic nanoparticles. *Sensors and Actuators B-Chemical* 107 (2): 936–944.

160 Ferreira, H.A., Graham, D.L., Feliciano, N. et al. (2005). Detection of cystic fibrosis related DNA targets using AC field focusing of magnetic labels and spin-valve sensors. *IEEE Transactions on Magnetics* 41 (10): 4140–4142.

161 Amaral, J., Cardoso, S., Freitas, P.P., and Sebastião, A.M. (2011). Toward a system to measure action potential on mice brain slices with local magnetoresistive probes. *Journal of Applied Physics* 109 (7): 07B308.

162 Pannetier, M., Fermon, C., Le Goff, G. et al. (2004). Femtotesla magnetic field measurement with magnetoresistive sensors. *Science* 304 (5677): 1648–1650.

163 Amaral, J., Pinto, V., Costa, T. et al. (2013). Integration of TMR sensors in silicon microneedles for magnetic measurements of neurons. *IEEE Transactions on Magnetics* 49 (7): 3512–3515.

164 Hamill, O.P., Marty, A., Neher, E. et al. (1981). Improved patch-clamp techniques for high-resolution current recording from cells and cell-free membrane patches. *Pflügers Archiv* 391 (2): 85–100.

165 Gullo, F., Maffezzoli, A., Dossi, E. et al. (2012). Classifying heterogeneity of spontaneous up-states: a method for revealing variations in firing probability, engaged neurons and Fano factor. *Journal of Neuroscience Methods* 203 (2): 407–417.

166 Ferrea, E., Maccione, A., Medrihan, L. et al. (2012). Large-scale, high-resolution electrophysiological imaging of field potentials in brain slices with microelectronic multielectrode arrays. *Frontiers in Neural Circuits* 6: 80.

167 Heer, F., Franks, W., Blau, A. et al. (2004). CMOS microelectrode array for the monitoring of electrogenic cells. *Biosensors and Bioelectronics* 20 (2): 358–366.

168 Baker, B.J., Kosmidis, E.K., Vucinic, D. et al. (2005). Imaging brain activity with voltage-and calcium-sensitive dyes. *Cellular and Molecular Neurobiology* 25 (2): 245–282.

169 Perron, A., Mutoh, H., Akemann, W. et al. (2009). Second and third generation voltage-sensitive fluorescent proteins for monitoring membrane potential. *Frontiers in Molecular Neuroscience* 2: 5.

170 Duan, X., Gao, R., Xie, P. et al. (2012). Intracellular recordings of action potentials by an extracellular nanoscale field-effect transistor. *Nature Nanotechnology* 7 (3): 174.

171 Xie, C., Lin, Z., Hanson, L. et al. (2012). Intracellular recording of action potentials by nanopillar electroporation. *Nature Nanotechnology* 7 (3): 185.

172 Hall, L.T., GCG, B., Thomas, E.A. et al. (2012). High spatial and temporal resolution wide-field imaging of neuron activity using quantum NV-diamond. *Scientific Reports* 2: 401.

173 Schotter, J., Kamp, P.B., Becker, A. et al. (2004). Comparison of a prototype magnetoresistive biosensor to standard fluorescent DNA detection. *Biosensors & Bioelectronics* 19 (10): 1149–1156.

174 Munoz-Rojas, F., Fernandez-Rossier, J., and Palacios, J.J. (2009). Giant magnetoresistance in ultrasmall graphene based devices. *Physical Review Letters* 102 (13): 136810.

4

Resonance Magnetometers

4.1 Introduction

The phenomena that referred to the notion of magnetic resonance are related to quantum radio-physics. Quantum radio-physics as a branch of science was formed in the beginning of the 1960s [1]. In this field of physics, one studies the phenomena accompanied with the emission or absorption of electromagnetic waves of radiofrequency diapason by quantum systems (nuclei, electrons, atoms, molecules, etc.). These phenomena, the physical nature of which is of independent interest, provided the basis of radio-spectroscopic methods for studying the structure of matter and physicochemical processes in it. They are also used for the creation of quantum generators, amplifiers, and magnetometers. Thus, the quantum radio-physics consists of radio-spectroscopy and radio-electronics. For the first time, the selective absorption of radio waves due to the magnetic properties of electron shells was observed in 1913 by Arkad'ev, Professor of Moscow University [2]. In 1934, Cleeton and Williams observed inversion spectra of ammonia in the range of radio frequencies [3]. In 1938, Rabi registered the first nuclear magnetic resonance (NMR) spectra in experiments with atomic beams [4]. The particularly intensive development of radio-spectroscopic methods began after 1944, when E. K. Zavoiskii, Professor of Kazan State University (Russian Federation), discovered the phenomenon of electron paramagnetic resonance (EPR) (Thesis for Doctor of Science, 1944) [1]. In 1946, two scientific teams of American scientists, headed by Bloch and Purcell, published articles on the observation of NMR in condensed matter [5, 6]. In 1950, Dehmelt and Krueger obtained spectra of nuclear quadrupole resonance (NQR) [7]. At present the latest technical achievements, including in the areas of processing and cryogenic technologies, are used in radio-spectroscopy. Devices developed on the basis of these achievements have extremely high sensitivity and resolution. Thus, the radio-spectroscopy allows

Magnetic Sensors for Biomedical Applications, First Edition. Hadi Heidari and Vahid Nabaei.
© 2020 by The Institute of Electrical and Electronics Engineers, Inc.
Published 2020 by John Wiley & Sons, Inc.

the investigation of processes that lead to the finest changes in the electronic structure of atoms and molecules.

Magnetic relaxation – the achievement of the equilibrium state of nuclear or electronic macroscopic magnetization in a static magnetic field – has no direct analogs in physical phenomena which determine spectra in infrared, visible, and more shortwave diapasons, when an equilibrium state is achieved mainly due to the spontaneous transitions from excited states. The magnetic relaxation processes are determined by the intensity of fluctuating electromagnetic fields in matter. Thus, the investigation of these processes is the study of interactions of nuclei and electrons with the fluctuating electromagnetic fields in close interweaving with the study of the nature and velocity of thermal molecular motion which produces those fluctuating fields. It should be noted that magnetic relaxation is sensitive to molecular motions in a very wide range. Most often, the study of spectral and relaxation characteristics of magnetic resonance have been held separately, and only in relatively rare original works attempts were made to carry out their complex analysis. However, the latter trend is increasing and no doubt in the nearest future these characteristics will be widely considered using common programs for the interpretation of responses of magnetic resonances at certain impacts on substances (or spin systems). The confirmation of this can be found in developing applications of double, ternary, and more complex resonances (2-, 3-, multidimensional spectroscopies), in which the relaxation processes are responsible for the transfer of spin polarization from one spin system to another.

The radio-spectroscopic methods are widely used in molecular physics, chemistry, biology, medicine, and other sciences. In recent years, the methods of quantum radio-physics are used in engineering and industry, as well as for control of technological processes. For example, in Russia, the NMR method is used for oil well logging, for the laboratory analysis of productivity oil collectors, and for the analysis of oil content and moisture in seeds. NMR system compatible with multi-type biological/chemical lab-on-a-chip assays has been reported in [8, 9]. The electron spin resonance (ESR) methods were developed for geological research, nondestructive testing of precious stones, etc. NMR and ESR have special characteristics with respect to galvanomagnetic sensors [10–12].

An important step in the formation of quantum radio-physics is connected with the publication of the articles of American physicists [13] and Soviet physicists [14], who had independently created the first quantum generators with a beam of ammonia molecules. Both groups noted that the initial purpose of their work was the creation of high-resolution radio-spectroscopes on the basis of molecular radiation, but much greater influence on the development of quantum radio-physics produced a "collateral" effect associated with the oscillating mode of the device, i.e. with its conversion in the quantum generator. Quantum generators with beams of atoms and molecules immediately attracted attention

due to their unique properties. The high frequency stability of quantum generators has determined their use as devices of storage precise time: time and frequency standards. Quantum amplifiers were created almost simultaneously
with quantum generators. The main advantage of quantum amplifiers is an
extremely low level of set (equivalent input) noises. The most perspective of
these were devices created on the basis of EPR in solids. In 1956, Bloemdergen
[15] showed that in systems with several energy levels it was possible to create a
steady-state quantum amplifier if one used an auxiliary radiation for the polarization of electron spins. It should be noted that for the first time the method of
inversion of level populations with pumping radiation was proposed by Basov
and Prokhorov [14]. Quantum amplifiers have provided new possibilities in
many areas of science and technology.

On the basis of quantum electronics devices magnetometers have been created, which have extremely high sensitivity and high accuracy measurements
of magnetic fields. Using quantum magnetometers, variations of the Earth's
magnetic field can be registered in limits of $10^{-7} \div 10^{-8}$, i.e. approximately
0.5×10^{-7} G (0.5×10^{-11} T). Quantum magnetometers are widely used for
the study of fluctuations of the Earth's magnetic field due to the change in conditions in the ionosphere and to internal processes in the globe depths, for
registration of spatial variations of the geomagnetic field in geological survey,
archeology, and military applications.

An important event that brought general recognition to quantum radio-
physics was the invention of lasers. In 1958, for the development of ideas of
quantum radio-electronics, Schawlow and Townes proposed a quantum generator of light and considered general problems related to the generation and
amplification of electromagnetic waves in the optical range [16]. The first pulsed
optical generator was created in 1960 [17] and the first steady-state one was realized in 1961 [18]. The appearance of coherent high-power light sources
resulted in a colossal leap in researches on nonlinear effects in physical optics,
as well as in many technical applications.

4.2 Nuclear Magnetic Resonance

As we might expect, for the observation of NMR, permanent magnetic field $\vec{B}_0$
is required, in which the Zeeman splitting of energy levels of the spin system
is realized, and an alternating radio frequency (RF) field $\vec{B}_1$, stimulating transitions between the Zeeman sublevels [1]. As a consequence of this interaction
the spin system can emit or absorb the energy of the RF field, if the frequency
of the field (ν) is close to the frequency of transitions (ν_0). Thus, for the observation of the NMR it is necessary to have the following main instruments:
(i) sources of static and alternating magnetic fields, (ii) a device that converts

the energy of the quantum transitions in the RF signals, (iii) amplifiers and hardware for registration of these signals.

Observation of NMR can be performed during continuous irradiation of alternating magnetic field $\vec{B}_1$, as well as between its pulses. In this regard, the principles of observation NMR in condensed media are divided into two groups, differing from each other both experimental methods and classes of solved tasks. In the first years after its discovery the NMR observed mainly with stationary RF exposure field $\vec{B}_1$, but soon the technique of its observations in the presence of pulsed RF field $\vec{B}_1$ was developed. Until the seventies, both techniques developed in parallels, solving the characteristic for their class specific tasks (the study of spectra on the basis of a stationary radiation field $\vec{B}_1$ and relaxation processes by pulsed irradiation). The development of experimental technologies and the emergence of powerful and high-speed PC led to the formation of the Fourier transform NMR spectroscopy, which united both directions.

4.2.1 Classical Model

Both classical and quantum mechanical viewpoints are integral parts of concepts in NMR spectroscopy. In the classical approach, the phenomenon of NMR is discussed in terms of net magnetization (M), in a macroscopic sample containing a large number of nuclei under study, rather than the behavior of individual spins [19]. In the presence of B_0, the angular momentum of the nucleus causes a precessional motion around the z-axis (Figure 4.1). The angular velocity of such a motion is given by $\omega_0 = -\gamma B_0$.

In the presence of B_0, more nuclei align in the parallel direction resulting in a net magnetization in the z direction (M_z). If an oscillating radiofrequency field (B_1) with a frequency equal to ω_0 is now applied in y-direction for a short time (t), then the net magnetization *nutates* (flips) from the z-direction towards x direction. The *flip-angle* due to such an RF pulse is given by $\theta = \gamma B_1 t$ (radians), where B_1 and t are the strength and the time duration, respectively, of the

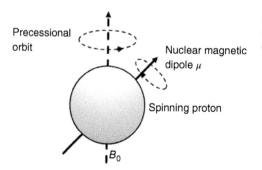

Precessional orbit

Nuclear magnetic dipole μ

Spinning proton

B_0

Figure 4.1 Larmor precession of a spin in magnetic field. *Source:* adapted from [19].

applied RF field. To start with, all spins precess with the same phase, after the application of RF field. This results in a transverse magnetization ($M_{x/y}$). In other words, the RF pulse creates a *phase coherence* among different precessing spins.

4.2.1.1 Rotating Frame of Reference

It is essential to understand the concept of rotating frame of reference, in which it is easier to follow the evolution of components of transverse magnetization (M_x and M_y). The rotating frame of reference is a coordinate system which rotates at the spectrometer frequency and is usually close to the individual Larmor frequencies of nuclear spins. In the rotating frame, M_x and M_y appear to precess with a frequency $\Omega_i = \omega_0(i) - \omega_{rf}$, where $\omega_0(i)$ are the individual Larmor frequencies and ω_{rf} is the applied RF (distinguished by the corresponding offset from the Larmor precession frequency).

It may be noted that Ω_i is of the order of kilohertz, as compared to ω_0, which is several megahertz. When $\omega_0(i) = \omega_{rf}$, $\Omega_i = 0$, the corresponding transverse magnetization appears stationary in the rotating frame. Such a reference frame helps to simplify several concepts in pulsed NMR. We will extensively use rotating frames. Hereafter, when the precessional frequencies are expressed by the symbol Ω, it is assumed that one is using the rotating frame of reference.

4.2.1.2 Strength of RF Pulses

In modern NMR techniques, RF pulses are characterized by their respective flip angles and pulse widths. For example, a $\pi/2$ pulse flips the net magnetization from the z-axis to the x–y plane and creates a transverse magnetization (Figure 4.2). Small flip angles rotate only a part of M on to the transverse plane. A π pulse inverts the magnetization to the negative z direction.

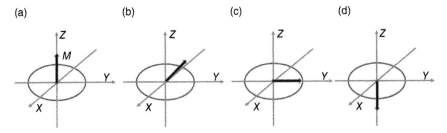

Figure 4.2 An RF pulse applied to the bulk sample creates a net magnetization in the transverse (xy) plane. In the rotating frame, the effect of the pulse is to tilt the net magnetization M which appears stationary. The flip angle depends on the strength and duration of the pulse. (a) The equilibrium magnetization. The net magnetization after the application of (b) $\pi/4$; (c) $\pi/2$; and (d) π pulse along x axis. In the laboratory frame, the net magnetization vector rotates with an angular velocity ω_0 around the longitudinal axis. *Source:* adapted from [19].

It may be pointed out that a $\pi/2$ RF pulse equalizes the population between the states α and β and hence there is no net z-magnetization. The transverse magnetization arises because of the *phase coherence* of the spins, which is created by the RF pulse. This coherence decays with time as spins go out of phase due to relaxation processes discussed later.

4.2.2 Basic Design of a NMR Spectrometer

The basic elements of a modern NMR spectrometer are shown in Figure 4.3. A large and highly homogeneous static magnetic field (B_0) is required to remove degeneracy among the nuclear-spin energy levels. The gap between the ground and the excited states increases with higher magnetic fields. Hence, the difference of populations of the spins between the two states is higher. Stronger magnetic fields help in achieving higher sensitivity. As discussed later, higher fields also help in a better dispersion of spectral lines. Thus, for studies on biological molecules, it is desirable to use the highest possible magnetic field strength. Spectrometers using field strengths as high as 21.15 T are currently available. This field corresponds to a ^{1}H resonance frequency of 900 MHz. It is also essential that the magnetic fields are homogeneous. Such high and homogeneous field strengths are produced using superconducting magnets. One often uses the ^{1}H resonance frequency to define the magnetic field used in a particular

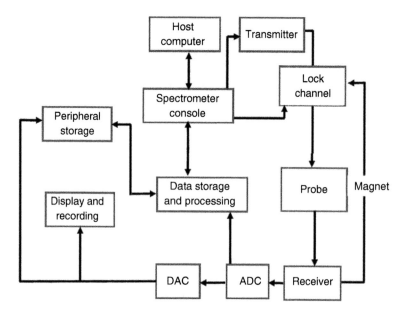

Figure 4.3 A schematic view of NMR spectrometer. *Source:* adapted from [19].

NMR spectrometer. Spectrometers are also distinguished by the magnet bore size. Magnets with narrow bores (typically a few centimeter), are used mostly for high-resolution solution work. The sample used in a high-resolution NMR experiment is generally around 600 µl and is taken typically in a sample tube of 5 mm in diameter. Wider bore magnets are used for studies on solid-state materials, which require additional hardware within the magnet bore such as the magic-angle spinning assembly and equipment for work over a wide temperature range. When working with humans and animals, lower magnetic field strengths are used because of safety considerations. RF penetration and heating issues and their associated neurological effects are important. The magnets should be able to accommodate humans or live animals. For animal studies, magnetic field strengths up to 7 T have been used. However, for humans, the optimum field strength for both MRI and MRS so far has been 1.5 T. In special cases 3 T magnets have been used. One of the important requirements of NMR experiments is a high homogeneity of the magnetic field, in the region of the sample (typically 1 in 10^9 over the entire sample volume). This is achieved through the use of shim coils, which provide additional fields to compensate for any variations. The shim coils are placed both in the cold part of the magnet and around the probe. A spinner assembly containing a small turbine system is used to spin the NMR sample tube to achieve better homogeneity.

The heart of the spectrometer is the probe. This has arrangement to hold the sample. Probe consists of various resonant circuits, and a set of RF coils and capacitors, needed for excitation and detection of the NMR free induction decay (FID) signal. At least three to four RF frequencies are needed to provide the ^{1}H frequency and one or two X-nucleus frequencies (e.g. for ^{13}C, ^{15}N, and ^{31}P). The magnetic field is locked to a master frequency called the deuterium lock.

For molecular studies, the RF probe head is usually introduced into the magnet from the bottom. In addition, there is provision to maintain sample at a constant temperature within $\pm 0.1\,^\circ$C. Most commercial spectrometers are equipped for variable temperature studies in the range -100 to $+150\,^\circ$C.

Magnetic field gradient coils are used to alter the static field. Such coils are incorporated either within the magnet assembly or surrounding the RF coil. Data accumulation over long time intervals requires a good field/frequency stability. An internal lock using ^{2}H of the solvent (^{2}H$_2$O) helps in compensating the drift in static magnetic field.

The console part of NMR equipment has a host computer which serves as a controller of the whole system and coordinates all the activities of the spectrometer. It has an arrangement for generating and transmitting RF at desired frequencies and power to the sample. Experiments are performed using RF fields (B_1) in the form of pulses. For multinuclear studies, RF fields at several frequencies are generated simultaneously. They are controlled, amplified, pulsed, and transmitted to the RF probe head. The RF field strengths are in the range of 0.01–0.2 mT, which correspond to several hundred kilowatts of power. RF pulses are programmed for diverse applications and are applied in a very precise time frame. Generation of different RF frequencies is achieved from a common master frequency derived from an oven packed crystal oscillator. Though RF pulses are applied for short durations, they can cause significant heating of the sample.

The detected NMR signal is an audio signal, which is digitized by an analogue-to-digital converter (ADC). ADC is characterized by its sampling rate (which is at least twice the frequency of the highest frequency signal that is to be detected; this is commonly known as *Nyquist criterion*) and the dynamic range (the ability of ADC to accurately measure signals having wide-ranging amplitudes). For optimum signal-to-noise (S/N) ratio, proper matching is provided between the excitation and the detection hardware. The detected signal which is a function of time (*time domain signal*) is fed into the computer memory. Later it is Fourier transformed (FT) and phase-corrected to get the frequency domain spectrum using either the host computer or an additional work-station. Software is available for acquiring, processing, displaying, plotting and basic analysis of the data on the host computer. This forms an important component of the spectrometer console.

4.2.3 Nuclear Magnetic Resonance in Molecular and Atomic Beams

The phenomenon of NMR for the first time was observed by I. Rabi and co-workers (1938), in experiments with atomic beams [1]. By this time, experiments of Gerlach and Stern [20] were already well known, which allowed to estimate the magnetic moment of an atom, as well as works of Estermann et al. [21] to determine the proton magnetic moment. All of them used the fact, that in an inhomogeneous magnetic field a particle, which has a magnetic moment ($\vec{\mu}$), is subjected by the force

$$\vec{F} = \mu_z \frac{\partial \vec{B}}{\partial z} \tag{4.1}$$

where $\partial \vec{B}/\partial z$ is the magnetic field gradient; μ_z is the component of the magnetic moment directed along the axis z. Under the influence of ($\vec{F}$) the particle will be deflected in the direction of the axis z. The magnitude of this deviation (dz) will be determined by the kinetic energy of the particle (E), its magnetic moment ($\vec{\mu}$) and magnetic field gradient:

$$dz = \vec{F} \frac{A}{E} = \mu_z \frac{\partial \vec{B}}{\partial z} \frac{A}{E} \tag{4.2}$$

A is a constant depending on the geometric parameters of the device.

Knowing instrument parameters, the value of $\partial \vec{B}/\partial z$ and the average kinetic energy of particles in a beam, as seen from Eq. (4.1), one can define μ_z from the value of dz. However, even in the most advanced experiments of Estermann et al. [21], the accuracy of the measurements of magnetic moments was not high: for the proton not better than 10%. For determining the magnetic moment of microparticles by the Rabi method, on a certain part of the particle trajectory, the conditions for magnetic resonance is created. The interaction of nuclear (or electronic) magnetic dipole with alternating field has been brought to its reorientation. The method consists in the following. The low-pressure chamber O (Figure 4.4a) ($P \cong 10^{-6}$ mm Hg $\cong 10^{-8}$ Pa) creates a beam of particles (atoms or molecules) which trajectories sequentially pass through the gaps of magnets A and B. These magnets are identical: inhomogeneous magnetic field in their gaps is equal in magnitude $|B_A|=|B_B|$, but with opposite direction of the vectors of field gradients, and $|\partial B/\partial z| \cong 800$ T/m. In this regard, the particle, flying through the gaps between magnets A and B, will be subjected to the same deflection, first, in one direction and then in the opposite direction and get on sensor detector D (Figure 4.4b).

If the investigated particles are molecules with compensated electron spins (molecules of $^1\Sigma_0$ – state) or atoms then the magnetic properties of these particles and their interaction with external magnetic fields are determined only by moments of the nuclei. Between the magnets A and B there is the magnet C, in

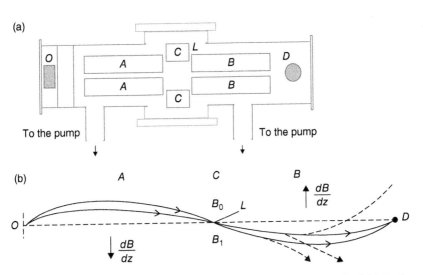

Figure 4.4 The functional diagram of the detection of NMR by Rabi method: (a) the low-pressure chamber O creates a beam of particles, which trajectories sequentially pass through the gaps of magnets A and B, and (b) principle of operation of the loop. *Source:* adapted from [1].

which gap a highly homogeneous magnetic field $\vec{B}_0$ is created. In addition, in the magnet gap the loop L of length l is set, which created an alternating magnetic field $\vec{B}_1(t)$, and which oriented perpendicular to the field $\vec{B}_0$. The principle of operation of the loop is illustrated in Figures 4.4b and 4.5a. As a result of effect of the static $\vec{B}_0$ and alternating $\vec{B}_1$ magnetic fields on magnetic dipole $\vec{\mu}$, this dipole will be re-oriented. The length of the loop l and the field strength B_1 is matched so, that for the span time (t) of particles through the magnet gap, all the dipoles $\vec{\mu}$ were reoriented (sign μ_z is changed to the opposite one).

Therefore, in accordance with Eq. (4.1), in the magnet gap B they will be deflected in the same direction as in the magnet gap A, and will not reach the detector D. Since the intensity of the reorientation process of magnetic dipoles increases with the frequency detuning ($\Delta\nu = \nu_0 - \nu$) decrease, then the flow particles falling into the slot of detector D must decrease. Consequently, the current i in the detector circuit will change simultaneously with the change $\Delta\nu$. This change of $i(\Delta\nu)$ has a resonant character. As an example, Figure 4.5b shows the NMR signal from nuclei ^1H in the beam of the KOH molecules, obtained in the $B_0 = 0.3453$ T.

The observed resonance lines in such a way allow us to determine the frequency and line width, to investigate the shape of the signal and other characteristics of the spectrum of NMR in atomic and molecular beams. If one studies hydrogen molecules, the NMR signal is observed only from the ortho-hydrogen (which possesses the total nuclear spin $I = 1$). In molecules of parahydrogen ($I = 0$) spins and magnetic moments of the proton pair are antiparallel, and the reorientation of the magnetic dipoles of individual protons in the magnet

(a)

(b)

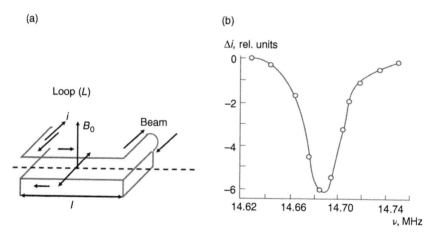

Figure 4.5 (a) The scheme of a loop for creating of the alternating magnetic field; (b) The resonance curve of ^1H nuclei in the beam of the KOH molecules. *Source:* adapted from [1].

C does not change the magnetic properties of the molecule as a whole and the trajectory of its movement. Using NMR on beams, gyromagnetic ratios and magnetic moments of many nuclei were measured with greater accuracy than it was done previously by other techniques.

4.2.4 The Sources of Magnetic Fields

The range of used fields values is approximately 0–20 T. As a rule in the region 0.1–2 T electromagnets or permanent magnets are used and fields with the induction of more than 2 T for the NMR can be obtained only in solenoids with a superconductive winding coil. At low fields (<0.1 T) one uses conventional solenoids or even Earth magnetic field (0.000 05 T).

The main requirements to the static magnetic field are (i) sufficiently high homogeneity within the scope of an investigated sample and (ii) field stability over time. The value of the average deviation of the field ΔB within a sample divided to the magnitude of the resonance field B_0 is taken as a measure of the field homogeneity. For measurements of relaxation times and self-diffusion coefficient in many cases it is sufficient to achieve the value $\Delta B/B_0$ of order of 10^{-5}–10^{-6}, but to obtain high resolution NMR spectra a value $\Delta B/B_0$ is required not worse than 10^{-8}–10^{-10}. A high degree of homogeneity in the gap of an electro-or permanent magnet can only be achieved through the set of design measures to gradually increasing the field homogeneity. The yoke of magnet is usually of H-shaped form that provides sufficient rigidity and symmetry of design in conditions of significant static compression loads between the poles (mutual force attraction of the pole pieces may be several tons).

To reduce edge effects (decreasing of the induction between poles at the edges of a gap relative to the center), the diameter of the pole pieces (D) is chosen much larger than the gap between poles (d): $D/d \geq 10$. With the same aim one can use the so-called *shimming* – a decrease of the d at the periphery of the gap by introducing additional annular rings (Figure 4.6a). By special design of the rings it is possible to expand significantly the volume of homogeneous field (see in Figure 4.6b, dashed curve). The shimming efficiency drops rapidly with an increase of the level of B_0 due to saturation of iron at $B_0 > 1$ T and in this connection this type of shimming is almost never used in modern instruments.

Pole pieces are made using soft (without carbon impurities) iron, followed by a thorough multi-stage machining, polishing and annealing in hydrogen atmosphere. As a result of processing, mechanical irregularities do not exceed 0.25 μm. Parallelism of surfaces of the pole pieces also significantly effects on the resulting homogeneity. Thus, the deviation from parallelism in 2 – 3 μm at $D \cong 30$ cm leads to inhomogeneity $\Delta B/B_0 \cong 10^{-6}$. Careful adjustment of poles with control by NMR signal allows us to obtain $\Delta B/B_0 \cong 10^{-7}$.

(a)

(b)

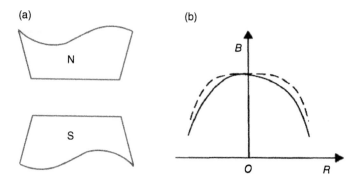

Figure 4.6 The influence of the shimming on the magnetic field in a magnetic gap. (a) Shimming is made by introducing additional annular rings and (b) by special design of the rings it is possible to expand significantly the volume of homogeneous field, see dashed curve. *Source:* adapted from [1].

4.2.5 NMR Spins Used in Life Science

All living systems have five major elements: carbon (C), hydrogen (H), nitrogen (N), oxygen (O), and phosphorous (P) [19]. Of these, the naturally abundant isotopes of C (^{12}C) and O (^{16}O) do not have nuclear magnetic moments. The more abundant isotope of N (^{14}N) has spin 1. Nuclei with spin >1/2 have quadrupole moments. As discussed later, such nuclei generally give rise to relatively broad signals with poor S/N ratio. Thus, one generally uses four nuclei in life sciences: ^{1}H, ^{13}C, ^{15}N, and ^{31}P, each of which has spin 1/2. These nuclear spins serve as reporter groups and provide complementary information on biological systems.

The individual NMR properties of above nuclei are listed in Table 4.1. Both hydrogen and carbon are present in all biological molecules and ^{1}H and ^{13}C can

Table 4.1 Important NMR active nuclear spins used in life sciences and their properties [19].

Nucleus[a]	Natural abundance	$\gamma(\times 10^6)$ (rads^{-1}T^{-1})	ν_0 in MHz at 11.743 T	Relative sensitivity	Chemical shift range (ppm)
^{1}H	99.985	267.522	500.000	100.0	15
^{2}H	0.015	41.066	76.753	0.009 65	20
^{13}C	1.11	67.283	125.725	0.02	250
^{15}N	0.37	−27.126	50.684	0.000 4	1 000
^{31}P	100.00	108.394	202.606	6.6	430

[a] All the listed nuclei possess spin(I) = 1/2, expect the ^{2}H whose I = 1.

provide information on all classes of biological functions. The high sensitivity of ^{1}H NMR coupled with its high natural occurrence makes it the most widely studied nucleus. Even ^{1}H resonance in a simple molecule like water (which constitutes almost 75% of all living systems) provides a wealth of information on biological systems. Nitrogen is an essential component of nucleic acids and proteins and ^{15}N provides useful information on such macromolecules. To avoid problems of sensitivity due to low natural abundance of ^{13}C or ^{15}N, one can use isotopically enriched (labeled) molecules. Like-wise, one can use ^{13}C or ^{15}N labeled substrates to trace metabolic pathways ^{31}P with its 100% natural abundance is another useful NMR probe. Though present in relatively smaller number of biological molecules, ^{31}P has been extensively used in the study of structure of nucleic acids and biological membranes. It has found wide applications for monitoring metabolism through molecules such as adenosine triphosphate (ATP) and phosphocreatine (PCr), which are involved in providing energy for essential processes of life.

While the above mentioned nuclei are those which are the most widely studied, use has also been made of isotopes such as ^{2}H ($I = 1$), ^{7}Li ($I = 3/2$), ^{14}N ($I = 1$), ^{19}F ($I = 1/2$), ^{23}Na ($I = 3/2$), for studying biological systems.

4.2.6 NMR Relaxation

Relaxation rates play an important role in NMR. Isolated nuclear spins do not have a mechanism to interact with the surroundings [19]. In the absence of suitable molecular interaction, it will take a very long time for the magnetization to relax back to their equilibrium values. Some of the methods by which nuclear spins can transfer their magnetization and dissipate energy to the surroundings are discussed below. These processes play an important role in NMR.

4.2.6.1 Relaxation Rates

To start with, when a sample containing spin 1/2 nuclei is subjected to magnetic field, the populations in the two levels α and β are equal. The net longitudinal magnetization $M_z = 0$ and the system are in a nonequilibrium state. To attain the equilibrium magnetization M_o, the system has to exchange energy with the surroundings. The build up of equilibrium magnetization follows a first-order differential equation (called the Bloch equation) with a typical rate constant R_1:

$$\frac{dM_z}{dt} = R_1(M_o - M_z) \tag{4.3}$$

The build-up or the decay of the longitudinal magnetization requires exchange of energy between the nuclear spins and the surroundings (*lattice*). This process is called longitudinal relaxation. The inverse of R_1 is called spin-lattice relaxation time or the longitudinal relaxation time (T_1). The rate at which the system attains equilibrium is governed by the mechanisms through which such an exchange of energy takes place.

Similarly, after excitation by a $\pi/2$ pulse there is no net magnetization in the z direction. The recovery of the longitudinal magnetization as a function of time (t) is mathematically described by:

$$M_z(t) = M_o\left[1 - \exp\left(-\frac{t}{T_1}\right)\right] \tag{4.4}$$

Once the spins have been excited by a $\pi/2$ RF pulse (i.e. the net magnetization has been nutated to the transverse plane), the coherent, precessing spins produce a detectable signal. The phase coherence or M_{xy} decays with a different time constant, T_2. This process is called spin–spin relaxation, because the energy for this process is exchanged through interaction between neighboring spins in the sample. The return to equilibrium is again governed by a simple first-order differential equation with a transverse relaxation rate $R_2(=1/T_2)$

$$\frac{dM_{xy}(t)}{dt} = -R_2 M_{xy}(t) \tag{4.5}$$

The solution to this differential equation is given by:

$$M_{xy}(t) = M_{xy}(0)\exp\left(-\frac{t}{T_2}\right) \tag{4.6}$$

For non-viscous liquids such as pure water, T_1 and T_2 are nearly equal. However, in general, the transverse magnetization decays faster than the longitudinal magnetization.

4.2.6.2 Molecular Mechanisms Leading to Relaxation

There are several mechanisms, whereby nuclear spins interact with electrons and other surrounding nuclear spins. Molecules constitute tiny magnets embedded in the form of nuclear magnetic moments and in some cases as unpaired electrons [19]. In a condensed state, the molecules undergo rotational and translational motions. This leads to fluctuations in the local magnetic fields at nuclear sites. Those frequency components in such fields, which exactly match the transition frequencies for the spin system under consideration, act as effective mechanisms for exchange of energy between the nuclear system and its surroundings. Thus, relaxation depends upon two factors:

1) a magnetic interaction, which is responsible for creation of the desired magnetic fields, and
2) molecular motions, which cause fluctuations in these fields.

Both intra- and intermolecular interactions lead to relaxation processes. Except for paramagnetic systems and quadrupolar nuclei, the major mechanism for energy exchange for spin 1/2 nuclei is the magnetic dipole–dipole (DD) interaction. For ^{1}H in molecules dissolved in aqueous solutions, a dominant relaxation mechanism is intermolecular DD interactions, since most of these nuclei are exposed to the solvent. On the other hand, ^{13}C spins are normally in the interior of the molecule and the relaxation is dominated by intramolecular ^{1}H – ^{13}C DD interactions. Other mechanisms for relaxation in diamagnetic samples include chemical shift anisotropy (CSA), scalar coupling (SC), and spin rotation (SR). In paramagnetic systems, interaction between the unpaired electron (e) and nuclear spins dominate relaxation mechanism rates. The magnetic moment of electron is several times larger than those of nuclei. Even a small amount of paramagnetic material (such as deoxyhemoglobin, dissolved oxygen, or paramagnetic metal ions) can strongly enhance relaxation rates. Nuclei with spin >1/2 relax mainly through electric quadrupole interaction (Q). Such nuclei also influence the relaxation of surrounding spins. It is possible to quantitatively define each of these interactions and estimate their relative contributions in relaxation processes. Thus, the relaxation rates (R) can be written as a sum of several contributions:

$$R = R(DD) + R(SR) + R(CSA) + R(SC) + R(Q) + R(e) \qquad (4.7)$$

4.2.7 NMR and Biological Structures

Applications of NMR to biological systems can be discussed at different levels of the complexity of biological structures [19]. The essential molecules involved in biological systems range from small molecules including water to complex and large macromolecules such as proteins and nucleic acids. Several of these molecules form multimolecular assemblies, such as protein–nucleic acid complexes, biological membranes, viruses, cell organelles. At the same time, a cell can be thought as the smallest unit of life. The cells are ultimately responsible for properties of organs in human, animals and plants. NMR has been applied to the whole range of biological complexity.

The knowledge of the 3D structures of proteins, t-RNA, ribosome, and their structure–function relationship is the next step in understanding various biological processes. Today, the 3D structures of a large number of such molecules are known. However, the available number is much smaller than what is needed for understanding the biochemistry of a human body. Several proteins with similar functions have similar 3D structures (examples are kinases, dehydrogenases). Therefore, knowledge of the structure of a particular class of protein from one organism may help to understand a particular biochemical reaction in a cell or organ of another organism. Thus, it is important to investigate all the unique protein folds that are involved in the function of living systems. This number of unique folds is estimated to be around 1000. The structure determination of proteins and nucleic acids and their *in-vitro* function thus forms an important area of biophysical research. NMR and X-ray crystallography are two major techniques used to unravel 3D structure of biological molecules.

The next level at which NMR has helped us to understand living systems is the macromolecular organization. An interesting application in this regard is the study of biological membranes, which show liquid crystalline behavior. Membrane bound proteins are often insoluble in water and are difficult to crystallize. Recent developments in high-resolution solid-state NMR techniques have made it possible to study such proteins. Cell function is a result of an intricate network of functioning proteins, small molecules and ions, which may activate or control enzyme functions. Even though attempts have been made to study enzyme function *in vitro*, one can understand *in vivo* biological functions by experimenting at cellular level. NMR has proved to be an invaluable tool to understand cell metabolism and in exploring factors which control and modify such actions.

Finally, in plants, animals and humans, different tissues or organs perform distinct functions. NMR is the technique of choice to understand such complex functions. Future strides in the methodologies of high-throughput NMR will further enhance our investigative capabilities to unravel complex biological structures and functions. Such studies are important for human health and understanding human physiology. A relatively new area is the development of NMR in plant and agricultural sciences.

4.2.8 Difficulties in Studying Biological System by NMR

It took almost 30 years for NMR to advance from chemical to biological systems. This is primarily because of several problems which one encounters when dealing with living systems. The solutions to these problems have led to major advances in applications of NMR to biology.

4.2.8.1 Sensitivity

By its very nature, NMR is a relatively insensitive technique. The intrinsic sensitivity of a nucleus is proportional to the cube of the nuclear magnetic moment. Since ^{1}H has the highest magnetic moment (μ), it is the most sensitive nucleus. This is fortunate because of the high abundance of ^{1}H in biological systems. The next most sensitive NMR spin is ^{19}F, which is less useful in biology. The natural abundance of the nucleus directly determines the number of nuclei available for detection of NMR signal. For example, natural abundance of ^{13}C is 1 in 100, while that of ^{15}N is 1 in 300. The sensitivity of less abundant nuclei however, can be increased by using labeled compounds.

The population difference between the ground and the excited site depends on the value of B_0. Use of higher fields therefore results in higher S/N. In the early days of NMR, solutions of 100 mM concentration of samples were required to obtain a decent ^{1}H NMR spectrum. It is almost impossible to obtain such high concentrations of biological macromolecules. With the advent of FT technology and advances in better RF probe designs, magnet and computer technology, this limit has now come down to sub-millimolar concentrations. One of the recent advances is the use of super-cooled RF-probes, with which one can enhance the S/N by an additional three to four times. In early 1960s, the S/N ratio recorded with a 100 mM sample of ethyl benzene was around 0.04 on a 30 MHz NMR spectrometer. Today, on a 900 MHz NMR spectrometer equipped with cryogenically cooled RF-probe one can achieve S/N of 8000, with 0.1% solution of ethyl benzene. This reflects a net gain of 200 000. However, in spite of these developments, the sensitivity of NMR experiments is much lower than that achieved by other spectroscopic techniques. In fact, the improvements in sensitivity are closely linked to the wider range of problems that can be handled by NMR.

4.2.8.2 Resolution
The ^{1}H signals lie in a narrow range of 0–15 ppm. Thus, it is difficult to resolve all the signals for larger molecules [19]. This is particularly true for spins in similar chemical environment, for example, the amide protons in proteins or the imino protons in nucleic acids. In the abovementioned example, one notices severe crowding of resonances for sugar ^{1}H resonances. One of the ways to handle such problem is to use higher fields for biomolecular studies. Higher fields also simplify the spectrum as they lead to better dispersion and convert strongly coupled spin systems to weakly coupled ones. The advent of multidimensional FT techniques has made further revolutionary advances in this regard, by expanding the information in more than one dimension.

4.2.8.3 Water Signal
With two protons, the molar concentration of water protons is almost 110 M in aqueous solutions. The signals from millimolar concentrations of molecules of interest have to be picked up against this background signal [19]. In most chemical applications, NMR studies are conducted using deuterated solvents. However, in biological applications studies are performed in H_2O, which is the milieu of living system. One can use ^2H_2O (D_2O) instead of H_2O to avoid strong water resonance for molecular studies. However, ^2H_2O should be avoided in animal studies. Even in studies on macromolecules, the use of ^2H_2O leads to H/D exchange for the labile NH and OH protons, which are important in studying hydrogen bonding. At the same time, the abundant presence of water in biological tissues, its strong signal, and sensitivity to its relaxation behavior due to its interaction with other molecules and its fluidity have been exploited in the development of powerful techniques such as magnetic resonance imaging (MRI).

4.2.8.4 Line Widths

The line widths of ^{1}H resonances in solutions of most organic compounds are less than 0.3 Hz. The lines in biological macromolecules are broader, because of restricted motion and small values of T_2. Line width increases with the molecular weight. Therefore studies on very large proteins are complicated not only by problems of resolution and assignments but also by the higher line widths. In cells and organs, the resonances are even broader. In recent years, methods to overcome such problems have been developed.

4.2.8.5 Quantification

The integrated area of the NMR signal is directly proportional to the concentration of the spin(s) giving rise to the signal [19]. The relaxation rates and therefore the line widths of NMR signals can be quite different depending on the molecular flexibility. Such differences are quite pronounced for the same molecule present in different parts of the cell or a living system. Line widths of even a simple molecule, such as water, may be resonances, may broaden below the noise level, and may not be detected. Even for the same molecule such as a biopolymer dissolved in aqueous solutions, different parts of the molecule may exhibit different motion and the line widths may not be comparable. This can create serious problem in quantification.

4.3 Magnetic Resonance Imaging

4.3.1 Introduction

The term *magnetic resonance imaging* (MRI), or else *magnetic resonance tomography*, involves a lot of techniques to get the image in a certain section of the spatial distribution of any characteristic of an object to which the magnetic resonance is sensitive. In a broader sense, the MRI means getting any localized information using magnetic resonance [1].

The idea to obtain localized NMR spectra and images with NMR was proposed a long time ago (Ivanov V. A. The method of determining the internal structure of objects. Auth. Certificate USSR, No. 1112266. Priority from 21.03.60. Bulletin of discoveries and inventions. 1984. No. 33. P. 125), but the first publication of the implementation of the method based on NMR appeared only in 1973. Shortly after this EPR images were obtained too. EPR imaging is still used as a laboratory research method (for example, to obtain information about the spatial distribution of free radicals), whereas NMR imaging (NMRI) for a long time went out beyond laboratories and it is widely used first of all for the purpose of medical diagnosis. Furthermore the NMRI is used in plant and animal physiology, geophysics, chemical technology, food industry, etc. Therefore, the physical principles of the localization of magnetic resonance signals will be considered in case of NMR. Let us consider the principles of establishing of the mutual correspondence between the NMR signal and the position of a volume element of an object (voxel) that produces this signal. There are many ways to establish such correspondence, but the basis of all of them is the proportionality between the magnetic resonance frequency and the magnitude of the magnetic field.

4.3.2 The Obtaining of Spin Images from NMR Induction Signals in Inhomogeneous Field

A common feature of all the methods of localization is that the experiment is performed in an inhomogeneous magnetic field with a specially organized character of inhomogeneity. If we observe a magnetic resonance in an inhomogeneous static magnetic field, different points in space correspond to the different resonance frequencies in an inhomogeneous field; hence space points are mapped by the frequency. However, the mapping because of the topological properties of the magnetic field is not biunique: the same frequency corresponds not to one, but to a lot of points in a space, located on certain surface. To overcome this difficulty the character of inhomogeneity is changed successively in the course of the experiment. The obtained thereby NMR signal from object contains the information on a spatial distribution of some parameter determining this signal, but this information appears in the implicit (encoded) appearance. Process and way of entering such an information in a magnetic resonance signal are termed as spatial encoding.

Decoding (deciphering) of a signal is produced by mathematical processing of the acquired data. Depending on the character of actions on spin system in the course of encoding, the visualization of spatial distribution of various parameters can be mapped, namely: concentrations of nuclei of the certain types or unpaired electrons, magnetic relaxation rates, chemical shifts, molecular diffusion, parameters of a macroscopic motion of a sample, etc. The EPR imaging is still typical stationary methods of magnetic resonance detection, but for obtaining NMR images, as well as in modern NMR spectroscopy, almost exclusively the pulse method is used. A variety of methods for producing spin images are divided into two main groups: the first one is *detection of free nuclear induction signal* in a nonuniform magnetic field, the second one is *excitation of NMR signals* in such a field.

To understand the essence of the first method of obtaining the spatial information, let us consider a simple experiment. Just after the action of 90° pulse (B_1 in Figure 4.7a), the magnetic field with a constant gradient (G_x in Figure 4.7b) is created in the region where the explored object is located. In this situation in each volume element termed *voxel*, there is the precession of nuclear magnetization with the frequency spotted by a voxel position, more exactly by value of its coordinate x. The observed signal of the FID from the whole object is the sum

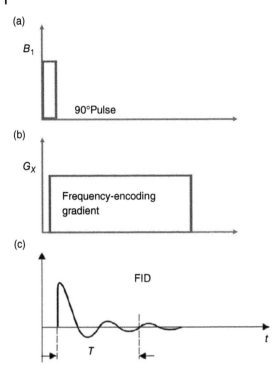

(a)

B_1

90°Pulse

(b)

G_x

Frequency-encoding
gradient

(c)

FID

t

T

Figure 4.7 Diagram of frequency-encoding pulse sequence with FID acquisition: (a) RF pulse; (b) frequency-encoding magnetic field gradient; (c) signal of free induction decay. *Source:* adapted from [1].

of signals from particular voxels in the object volume (Figure 4.7c). Hence, the Fourier spectrum of this signal with an appropriate choice of the scale represents the distribution of density of resonating spins along the coordinate x. In other words, the frequency spectrum of the signal of the magnetic resonance is the projection of density distribution of resonating spins on the direction of gradient of a magnetic field.

4.3.3 MRI Instrumentation

This section discusses basic instrumentation aspects as they relate to the MRI scanner and its functionality. These include the magnet and the various designs, matters on field homogeneity, stability, shimming, and the fringe field, gradient coils and RF coils and their designs, other electronics (transmitters, receivers), and decoupling schemes. A general overview of the scanner and its auxiliary systems is shown in Figure 4.8.

Specifically, and in a block diagram form, the MRI system and the RF chains (transmitter, receiver, modulator, and receiver demodulator) are depicted in Figure 4.9. In particular, the transmitter-receiver, the transmitter modulator,

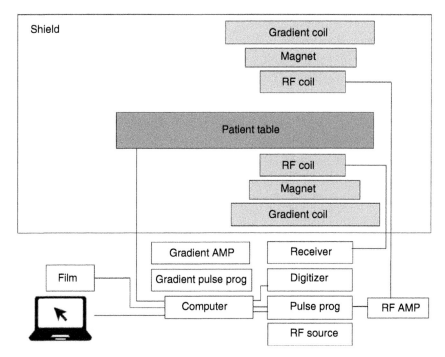

Figure 4.8 A schematic representation of all MR hardware and interconnections. *Source:* adapted from [22].

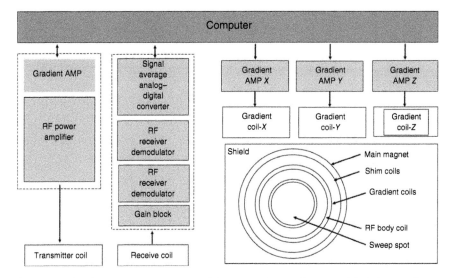

Figure 4.9 A schematic representation of the transmitter and receiver RF chains, gradient coils, and main magnet room. *Source:* adapted from [22].

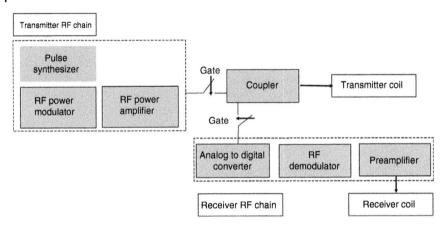

Figure 4.10 A schematic representation of the transmitter and receiver RF chains. Gates are used to isolate the receiver from the transmitter (during the receive phase and vice versa). Such isolation reduces coupling and noise. *Source:* adapted from [22].

and the receiver demodulator are drawn in Figure 4.10 in a more detailed block diagram form. Once the signal is received and demodulated to a frequency of a few kilohertz, a double-balanced demodulator is used and the signal is detected in quadrature. It is then sampled and inverse FT for reconstruction (Figure 4.11).

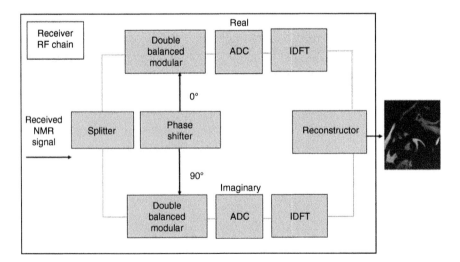

Figure 4.11 A schematic representation of the receiver RF chain and quadrature reconstructor. *Source:* adapted from [22].

4.3.3.1 Magnets and Designs

One of the most important criteria for superior performance and image quality is a highly uniform magnetic field over the imaging volume (typically of the order of 40 – 50 cm^3). To allow such homogeneity, while providing at the same time easy patient access, comfort of the patient, and minimal claustrophobia, magnets have endorsed designs of a cylindrical shape with a hollow cylindrical bore. Modern open and portable MRI systems have recently become available; however, they still lack somewhat in homogeneity performance with respect to closed-magnet (open-bore) systems. The magnetic field function that is produced along the direction of the z axis can be expressed in terms of a Taylor series expansion as a summation of the first field harmonic and higher-order terms that involve Legendre polynomials of degree n. In general the magnetic field at the center of the Cartesian axes (at $z = 0$) can be written as

$$B_z = J \cdot L \cdot R_0 F(\chi_1, \chi_2, \chi_3, ..., \chi_n) \tag{4.8}$$

where J is the current density in the coil conductor (current per unit area), L is the ratio of conductor cross-sectional area to the total area, R_0 is the inner radius of one of the magnet coils, and F is a filling factor function that depends on a number of geometrical variables χ_1. The function F can be expanded using a Taylor series approximation of about $z = 0$, and the axial field can be written as

$$B_z = JLR_0 \left[1 + e_2 \left(\frac{z}{R_0} \right)^2 + e_4 \left(\frac{z}{R_0} \right)^4 + ... \right] \tag{4.9}$$

where the coefficients e_n are dimensionless and geometry dependent. The odd terms in the expansion vanish by symmetry. Field homogeneity is optimum when the maximum number of low-order coefficients, e_n, vanishes. A magnet is described as a magnet with an order ith if the e_i coefficient vanishes. So, a Helmholtz pair is fourth order if e_4 vanishes, or equivalently, the expansion of B_z has terms up to and including e_3. Most commercial scanners are of order 6 or 8, and the requirement for higher field homogeneity is achieved with the use of additional shim coils (Figure 4.9) that through current flow adjustments (which are computer controlled) can achieve superior performance and field homogeneity.

4.3.3.2 Resistive Electromagnets

Most systems are built around a superconducting magnet or an electromagnet. Both types use electrical current to generate a uniform field. An adequate field is produced by several pairs of large-diameter coils of wire known as Helmholtz coils. The primary field is made as homogeneous as possible by moving about shim plates of steel (passive shimming) and by adjusting the small currents in various additional shim coils, which produce slight field corrections (active shimming).

So for an electromagnet, copper or aluminum conductors are wound on an aluminum frame with a bore diameter of about 1 m within which the gradient field coils and RF coils are fitted. An inner bore of about 40–60 cm is left for the patient. Conductor Joule heating occurs so the current-carrying conductors are hollow, to allow circulating cooling water to cool them down. Resistive magnets can attain field strengths of about 0.01–0.2 T.

4.3.3.3 Permanent Magnets

Large blocks of ferrous metal are used to generate the magnetic field. Unlike the other types, the main field of the permanent magnet is aligned vertically. Such magnets are largely maintenance-free and consume no electric power or cryogens (i.e. liquid helium and nitrogen), but their disadvantage is that they are capable of producing only relatively low fields (up to approximately 0.4 T). Two basic designs, the ring dipole and the H-frame, are shown in Figure 4.12. Eight segments compose the ring dipole, with the magnetization indicated by the arrows. The H-frame consists of two massive pole pieces held together by a steel frame. Although bulkier in size and structure, the H-frame design is preferred due to improved fringe field advantages.

4.3.3.4 Superconducting

Most modern machines use a superconductor. A superconducting magnet exploits the significant reduction of electrical resistance in some materials at very low temperatures (e.g. niobium-titanium alloy at −253 °C). Once the current is initiated, there is no further need for power input, nor are there any conducting losses.

To maintain the superconducting condition, the entire coil must be immersed in liquid helium (approximately −269 °C) contained within a cryostat (helium Dewar vessel), i.e. a stainless steel vacuum insulated cryostat that is filled with liquid helium. A surrounding refrigerator prevents warming of helium from the

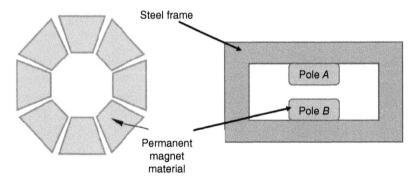

Figure 4.12 Types of permanent magnets. *Source:* adapted from [22].

outside sources. Some systems use liquid nitrogen instead of a refrigerator system (boiling point of approximately −196 °C) to prevent heat inflow to the helium container. Helium boils off slowly, so it has to be replenished every few months, and so does liquid nitrogen (even more frequently). The high cost associated with the MRI unit is mostly due to the manufacture of the magnet-cryogen assembly and its cryogenic power consumption. In the United States, the Food and Drug Administration (FDA), currently limits clinical diagnostic studies on humans in magnets with a field strength of at most 3 T (i.e. 30 000 G).

4.3.3.5 Stability, Homogeneity, and Fringe Field

A number of important features of the magnetic field and the magnet type itself, including the temporal field stability, field homogeneity, and fringe field, are critical for the image spatial resolution, and ultimate signal-to-noise ratio (SNR) performance. Temporal stability is determined by the magnet type and design and by external time-varying fields and interactions. Resistive magnets, for example, are highly dependent on power variations. Temperature fluctuations are also extremely important for permanent magnets. Superconducting magnets are the most stable, with an expected resonant frequency shift at 1.5 T of less than 1 MHz in approximately 20 years. Homogeneity is also a critical determinant of the image spatial resolution.

For clinical systems, field homogeneity is expected to be in the range of approximately 15 ppm within the central 40 cm diameter spherical volume. Within the "sweet spot" the FWHM of a line width must be less than about 0.3–1.0 ppm. Magnetic shielding is often used to reduce the fringe field, that is, the field extending outside the magnet room. Active or passive shielding is often achieved with specialized coils generating a field to counteract the fringe field extent, or alternatively with a ferromagnetic passive shield enclosure.

4.3.3.6 Gradient Coils

Gradient coils are used to generate the linear, spatially varying weak magnetic fields necessary for spatial encoding. Their design is based on use of multiple current-carrying conducting loops, similar to the designs of the main magnet types. A number of techniques are used to optimize the design of such coils, based on inductance, power dissipation, and field linearity. Both the gradient coils and the RF coils can use the Biot–Savart law to compute the magnetic field distribution. For a magnetic field in three-dimensional space $B(r)$, generated by a unit length of a conductor δl, carrying a current I, Biot–Savart states that for any point in space $P(x, y, z)$ the magnetic field is given by

$$B(r) = \frac{\mu_0 I}{4\pi} \oint \frac{dl \times (r - r')}{|r - r'|^3} dr \qquad (4.10)$$

where μ_0 is the magnetic permeability and r and r' are the distances of the conducting wire and the point in space for which the magnetic field is being computed, respectively. In modern clinical scanners, gradient windings are wrapped around a cylindrical former in patterns that optimize linearity and minimize power consumption. Typical examples of axial gradient coils include the Maxwell pair and the Golay coils.

4.3.3.7 RF Coils

RF coils are used for signal reception from the magnetization in the transverse plane. RF coils are simply antennas or, alternatively, resonant LC circuits tuned at the Larmor resonant frequency. The resonant frequency is determined by

$$f_o = \frac{1}{2\pi\sqrt{LC}} \tag{4.11}$$

Another important and critical factor of their performance is the quality factor Q, often defined as the 3 dB bandwidth of their frequency response:

$$Q = \frac{f_o}{f \pm 3\text{dB}} \tag{4.12}$$

RF coils are designed to receive the weak RF signal from within the human body. They are classified into transmit and receive coils. Although on certain occasions a single coil is used to transmit and receive the MRI signal (such as the body coil that is built in most scanner systems), often separate transmit and receive coils are used with connections to the transmitter and receiver RF chains, as shown in Figure 4.10. Three major types of RF coils have been introduced in the MRI field: *surface, volume,* and other specialized coils known as the *phased arrays.* Matters of importance regarding surface coils are their designs, tuning, and matching, their interconnections to the scanner, their decoupling during RF transmission and reception, and their safety. The design of surface coils is also an active area of MRI research, and numerous coil specialized designs exist.

4.3.3.8 RF Decoupling

A critical and most important issue in the use of separate transmit and receive coils is their electrical isolation and decoupling during the two phases of the MRI experiment, that is, transmission of RF power and reception of the MRI signal. Usually, passive, high-breakdown PIN diode switches are used in parallel configurations on the RF transmit coils. Induction of voltages during the transmit phase (which can be as large as a few hundred volts) forward biases the diodes. During the reception of the weak RF signal the diodes are reverse biased and the transmitter coil effectively becomes an open circuit. More elaborate active biasing schemes are often used with a DC (direct current) voltage provided from a power supply, usually residing in the receiver cabinet. Such a scheme is used for the protection of receiver coils. The example in Figure 4.13 shows a resonant loop circuit in a phased array with blocking circuits on the upper and lower right ends that include a PIN diode. The diodes are biased by a DC voltage during the transmit phase. This introduces two separate resonant circuits within the resonant loop structure that causes splitting of the resonant peak at a frequency other than the transmitter frequency. Other important connecting parts are the balun circuits often used to reduce or eliminate circulating currents, thereby providing protection to the patients (Figure 4.13).

4.3.4 MRI of Flow

The effect of flow on the observed NMR signal was first described by Suryan [23] soon after the discovery of NMR [24]. Suryan observed an increased NMR signal from a flowing liquid as compared to the signal from the same liquid at rest. Suryan correctly explained the observed effect by the decreased saturation of the magnetization in the flowing liquid. A decade later the same effect was reported by Singer [25, 26] who, along with Bowman and Kudravcev [27], developed several techniques for quantitative NMR flow measurements. The studies of these and other investigators laid the groundwork for MR flow imaging. Reference [28] contains an interesting review of early NMR studies of flow.

After flow-related effects in MR images were first discussed by Young et al. [29] and Crooks et al. [30] in the early 1980s, various techniques have been developed for MRI of flow. In general, these techniques can be divided into three major categories based on the effects used for flow imaging: (i) "time-of-flight" effects [31–44]; (ii) velocity-induced phase [45, 46]; and (iii) signal enhancement caused by MR contrast agents [47–52].

4.3.4.1 Time-of-Flight Techniques
4.3.4.1.1 Gradient-Echo Imaging
Spatially selective r.f. pulses are used in 2D and 3D MR imaging to repeatedly excite magnetization in a slice of material [24]. When flowing spins enter an excited slice during imaging, they normally have a higher longitudinal magnetization than that of the saturated static material in the slice. Therefore, in-flow of unsaturated spins makes the flowing material in images appear brighter than the static material. Time-of-flight (TOF) MR imaging utilizes the difference in magnetization of stationary and flowing spins in order to improve visualization of flow in MR images. In 2D TOF imaging the signals are acquired from consecutively excited thin slices (1.5–3 mm thickness), which are typically chosen perpendicular to the direction of flow in order to minimize saturation of the flowing material. Because it is often difficult to assess flow conditions from conventional two-dimensional MR images, additional images of flow can be produced by using a post-processing technique, known as the *maximum*

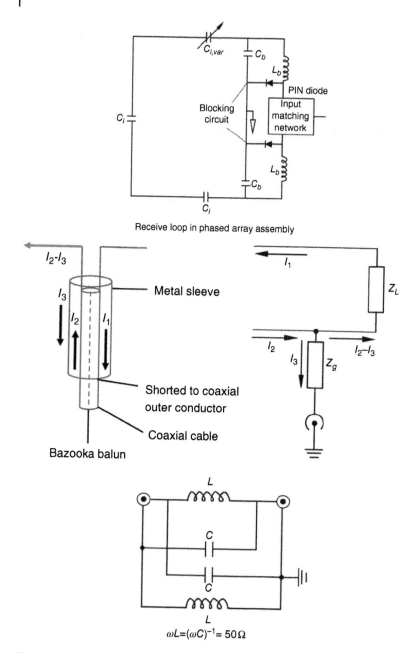

Receive loop in phased array assembly

$\omega L = (\omega C)^{-1} = 50\,\Omega$

Figure 4.13 Top row: Blocking resonant circuitries (L_b, C_b) on the resonant coil with active decoupling activated upon DC bias of the PIN diodes. Middle row: A bazooka balun circuit at the connecting end of the RF antenna to minimize circulating currents. Bottom row: An alternative balun circuit using lump components. *Source:* adapted from [22].

intensity projection (MIP). In this technique an imaged volume is projected onto a plane. The determined maximum voxel intensities along rays perpendicular to the projection plane are displayed subsequently as two-dimensional MIP images. The MIP algorithm is very effective for selectively displaying high-intensity regions in the imaged volume.

Since 3D imaging in general provides higher SNR and allows better resolution than 2D imaging, it is often chosen as the technique for diagnostic MR imaging of arteries, known as *magnetic resonance angiography* (MRA). Because in 3D imaging signals are acquired following excitation of a relatively thick slice (slab), the flowing material (blood) is continuously saturated by successive excitation pulses as it propagates through the slab. This saturation effect is particularly significant under the conditions of slow flow or flow traveling parallel to the imaged slab. The saturation of blood is spatially dependent, varying from its minimum near the entrance to the slab to its maximum near the exit. Several imaging techniques have recently been suggested in order to ameliorate contrast in 3D TOF MRA. One of these techniques employs r.f. excitation with a spatially varying flip angle that increases from its minimum value at the slab entrance to its maximum value at the slab exit [36]. In another approach saturation of flowing material in 3D TOF MRA is decreased by sequential imaging of overlapping thin slabs [37].

In many instances the close proximity of veins to arteries might make it difficult to differentiate between the two in MR images. Visualization of arterial structures in TOF MRA can be improved by selectively saturating the magnetization of venous blood. The use of selective saturation in TOF MRA is facilitated by the fact that the direction of arterial blood flow is typically opposite to that of venous blood flow. In practice the magnetization of venous blood is reduced routinely by selectively saturating a slab of tissue through which the blood flows prior to entering the imaged volume. The resulting effect is the dark appearance of veins relative to the bright appearance of arteries in TOF images.

4.3.4.1.2 Spin Echo Imaging

Initially excited blood remains in the imaging volume during the time $\tau = d/v$, where d is the slice thickness and v is the component of the flow velocity perpendicular to the slice. If the time interval $TE/2$ between excitation and the spatially selective 180° pulse in spin-echo imaging exceeds τ, then only the magnetization of the static material is refocused. Therefore, the washout of flowing spins from the imaged slice, known as the *out-flow effect*, makes blood vessels appear darker than the surrounding tissue in the image. For example, spin-echo images acquired with slice thickness of 2 mm, TE of 20 ms would display a greatly reduced signal from flow (flow void) if the flow velocity exceeds 20 cm/s. Imaging techniques based on the outflow effect [38] are frequently referred to as "black blood" MRA because of the darker appearance of blood relative to that of stationary material in the MRA images. Conversely, techniques used to acquire images with bright appearance of blood are known as "bright blood" MRA. Unlike bright blood MRA, black blood MRA techniques generate flow contrast by minimizing the signal from moving spins. As a result, these techniques also minimize artefacts due to blood flow pulsations, saturating of the blood magnetization and flow-induced dephasing of spins which are often present in bright blood MRA images.

4.3.4.1.3 Inversion-Recovery Imaging

To improve contrast between flowing and stationary spins, Nishimura et al. [39] suggested a subtraction technique that can completely eliminate signal from the static material. In this technique two images are acquired as follows: the first image is obtained following a 180° pulse that selectively inverts magnetization of the spins in the slab and above; the second image is acquired after the magnetization in the entire volume is inverted (Figure 4.14). In both cases the same time delay, TI, is used between the inversion and excitation pulses. Image subtraction leads to cancellation of the signals from both moving and stationary spins that have experienced inversion pulses. The resulting image is characterized by a bright appearance of vessels with upstream flow entering the slab from below. The major advantage of this approach as compared to the gradient-echo techniques described earlier is potentially better visualization of vessels with slow flow, provided that the inversion time TI is long enough so that complete replenishment of the flowing material in the slab takes place. The major limitations are long scan time, because two images are acquired, and increased sensitivity to patient motion during data acquisition, because image subtraction is used.

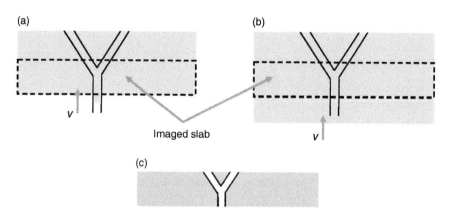

Figure 4.14 A diagram describing selective-inversion MRA with image subtraction: (a) first image is acquired following inversion of the magnetization in the slab and above; (b) second image is acquired after magnetization in the entire volume is inverted; (c) image subtraction results in cancellation of the signal from the static material in the imaged slab.
Source: adapted from [24].

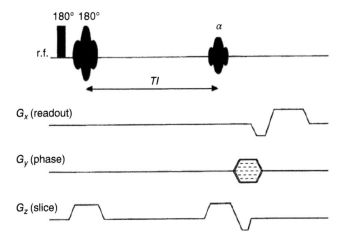

Figure 4.15 Pulse-sequence for selective-inversion black blood MR imaging. *Source:* adapted from [24].

In another approach selective inversion of the longitudinal magnetization is used in order to null the signal from blood [40]. In this black blood MRA approach, a nonselective 180° pulse first inverts the longitudinal magnetization in the entire volume. Immediately after the first inversion a spatially selective 180° pulse returns the magnetization to its equilibrium state in the imaged slice only (Figure 4.15). During the inversion time, *TI*, the blood that initially experienced selective 180° pulse leaves the slice. The longitudinal magnetization of the blood entering the slice at time *TI* is proportional to $1 - 2e^{-TI/T_{1,blood}}$, where $T_{1,blood}$ is the spin-lattice relaxation time of blood.

4.4 Electron Spin Resonance

ESR spectroscopy, also known as EPR or electron magnetic resonance (EMR), was invented by the Russian physicist Zavoisky in 1945 [53]. It was extended by a group of physicists at Oxford University in the next decade. Reviews of the Oxford group's successes are available [54–56] and books by Abragam and Bleaney and by Abragam [57, 58] cover the major points discovered by the Oxford group. In the present book, we focus on the spectra of organic and organotransition metal radicals and coordination complexes. Although ESR spectroscopy is supposed to be a mature field with a fully developed theory, there have been some surprises as organometallic problems have explored new domains in ESR parameter space. We will start in this chapter with a synopsis of the fundamentals of ESR spectroscopy. For further details on the theory and

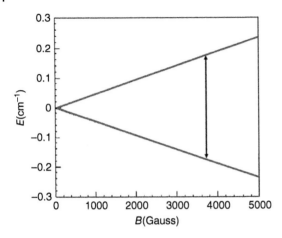

Figure 4.16 Energy levels of an electron placed in a magnetic field. *Source:* adapted from [69].

practice of ESR spectroscopy, the reader is referred to one of the excellent texts and monographs on ESR spectroscopy [59–67]. The history of ESR has also been described by many of those involved in the founding and development of the field [68].

The ESR spectrum of a free radical or coordination complex with one unpaired electron is the simplest of all forms of spectroscopy. The degeneracy of the electron spin states characterized by the quantum number, $m_s = \pm 1/2$, is lifted by the application of a magnetic field, and transitions between the spin levels are induced by radiation of the appropriate frequency (Figure 4.16). If unpaired electrons in radicals were indistinguishable from free electrons, the only information content of an ESR spectrum would be the integrated intensity, proportional to the radical concentration. Fortunately, an unpaired electron interacts with its environment, and the details of ESR spectra depend on the nature of those interactions. The arrow in Figure 4.16 shows the transitions induced by 0.315 cm^{-1} radiation.

Two kinds of environmental interactions are commonly important in the ESR spectrum of a free radical: (i) To the extent that the unpaired electron has residual, or unquenched, orbital angular momentum, the total magnetic moment is different from the spin-only moment (either larger or smaller, depending on how the angular momentum vectors couple). It is customary to lump the orbital and spin angular momenta together in an effective spin and to treat the effect as a shift in the energy of the spin transition. (ii) The electron spin energy levels are split by interaction with nuclear magnetic moments – the nuclear hyperfine interaction. Each nucleus of spin I splits the electron spin levels into $(2I + 1)$ sublevels. Since transitions are observed between sublevels with the same values of m_I, nuclear spin splitting of energy levels is mirrored by splitting of the resonance line (Figure 4.17).

Figure 4.17 Energy levels of an unpaired electron in a magnetic field interacting with a spin-1/2 nucleus. The arrows show two allowed transitions. *Source:* adapted from [69].

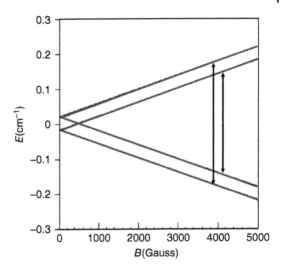

4.4.1 The ESR Experiment

When an electron is placed in a magnetic field, the degeneracy of the electron spin energy levels is lifted as shown in Figure 4.16 and as described by the spin Hamiltonian:

$$\hat{H}_s = g\mu_B B \hat{S}_z \tag{4.13}$$

where, g is called the g-value (or g-factor), ($g_e = 2.002\,32$ for a free electron), μ_B is the Bohr magneton (9.274×10^{-28} J/G), B is the magnetic field strength in Gauss, and S_z is the z-component of the spin angular momentum operator (the magnetic field defines the z-direction). The electron spin energy levels are easily found by application of $\hat{H}_s$ to the electron spin eigenfunctions corresponding to $m_s = \pm 1/2$:

$$\hat{H}_s \left\langle \left| \pm \frac{1}{2} \right\rangle = \pm \frac{1}{2} g\mu_B B \left\langle \left| \pm \frac{1}{2} \right\rangle = E_{\pm} \left\langle \left| \pm \frac{1}{2} \right\rangle \right. \tag{4.14}$$

Thus

$$E_{\pm} = \pm \frac{1}{2} g\mu_B B \tag{4.15}$$

The difference in energy between the two levels

$$\Delta E = E_+ - E_- = g\mu_B B = h\nu \tag{4.16}$$

corresponds to the energy, $h\nu$, of a photon required to cause a transition; or in wavenumbers by Eq. (4.17), where $g_e \mu_B / hc = 0.9348 \times 10^{-4}$ cm^{-1}G^{-1}:

$$\tilde{\nu} = \lambda^{-1} = \frac{\nu}{c} = \left(\frac{g_e \mu_B}{hc} \right) B \tag{4.17}$$

Since the g-values of organic and organometallic free radicals are usually in the range 1.9–2.1, the free electron value is a good starting point for describing the experiment. Magnetic fields of up to c.15 000 G are easily obtained with an iron-core electromagnet; thus we could use radiation with $\tilde{v}$ up to 1.4 cm^{-1} ($v <$ 42 GHz or $\lambda > 0.71$ cm). Radiation with this kind of wavelength is in the microwave region. Microwaves are normally handled using waveguides designed to transmit radiation over a relatively narrow frequency range. Waveguides look like rectangular cross-section pipes with dimensions on the order of the wavelength to be transmitted. As a practical matter for ESR, waveguides cannot be too big or too small –1 cm is a bit small and 10 cm a bit large; the most common choice, called X-band microwaves, has λ in the range 3.0–3.3 cm ($v \approx 9 - 10$ GHz); in the middle of X-band, the free electron resonance is found at 3390 G. Although X-band is by far the most common, ESR spectrometers are available commercially or have been custom built in several frequency ranges (Table 4.2).

4.4.1.1 Sensitivity

As for any quantum mechanical system interacting with electromagnetic radiation, a photon can induce either absorption or emission. The experiment detects net absorption, i.e. the difference between the number of photons absorbed and the number emitted. Since absorption is proportional to the number of spins in the lower level and emission is proportional to the number of spins in the upper level. ESR sensitivity (net absorption) increases with the total number of spins, with decreasing temperature and with increasing magnetic field strength. Since the field at which absorption occurs is proportional to microwave frequency, in principle sensitivity should be greater for higher frequency K- or Q-band spectrometers than for X-band. However, the K- or Q-band waveguides are smaller, so samples are also necessarily smaller and

Table 4.2 Common frequencies used for ESR (available commercially or have been custom built) [69].

Designation	v (GHz)	λ (cm)	B (electron) (G)
L	1.1	27	390
S	3.0	10	1 070
X	9.5	3.2	3 400
K	24	1.2	8 600
Q	35	0.85	12 500
W	95	0.31	34 000
—	360	0.083	128 000

for the same concentration contain fewer spins. This usually more than cancels the advantage of a more favorable Boltzmann factor for samples of unlimited size or fixed concentration.

Under ideal conditions, a commercial X-band spectrometer can detect about 10^{12} spins (c. 10^{-12} moles) at room temperature. This number of spins in a 1 cm^3 sample corresponds to a concentration of about 10^{-9} M. By ideal conditions, we mean a single line, on the order of 0.1 G wide, with sensitivity going down roughly as the reciprocal square of the line width. When the resonance is split into two or more hyperfine lines, sensitivity decreases still further. Nonetheless, ESR is a remarkably sensitive technique, especially compared with NMR.

4.4.1.2 Saturation

Because the two spin levels are affected primarily by magnetic forces, which are weaker than the electric forces responsible for most other types of spectroscopy, once the populations are disturbed by radiation it takes longer for equilibrium population differences to be established. Therefore an intense radiation field, which tends to equalize the populations, leads to a decrease in net absorption which is not instantly restored once the radiation is removed. This effect is called "saturation." The return of the spin system to thermal equilibrium, via energy transfer to the surroundings, is a rate process called spin–lattice relaxation, with a characteristic time (T_1), the spin–lattice relaxation time (relaxation rate constant = $1/T_1$). Systems with a long T_1 (i.e. spin systems weakly coupled to the surroundings) will be easily saturated; those with shorter T_1 will be more difficult to saturate. Since spin–orbit coupling provides an important energy transfer mechanism, we usually find that odd-electron species with light atoms (e.g. organic radicals) have long T_1^s, those with heavier atoms (e.g. organotransition metal radicals) have shorter T_1^s.

4.4.2 Operation of an ESR Spectrometer

Although many spectrometer designs have been produced over the years, the vast majority of laboratory instruments are based on the simplified block diagram shown in Figure 4.18. Plane-polarized microwaves are generated by the klystron tube and the power level adjusted with the Attenuator. The Circulator behaves like a traffic circle: microwaves entering from the klystron are routed toward the Cavity where the sample is mounted. Microwaves reflected back from the Cavity (which is reduced when power is being absorbed) are routed to the Diode Detector, and any power reflected from the diode is absorbed completely by the Load. The diode is mounted along the E-vector of the plane-polarized microwaves and thus produces a current proportional to the microwave power reflected from the cavity. Thus, in principle, the absorption of microwaves by the sample could be detected by noting a decrease in current in the Microammeter. In practice, of course, such a

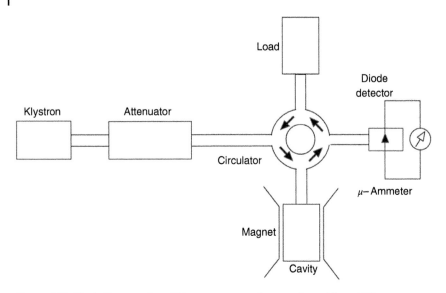

Figure 4.18 Block diagram of an ESR spectrometer. *Source:* adapted from [69].

measurement would detect noise at all frequencies as well as signal and have a far too low SNR to be useful.

The solution to the signal-to-noise problem is to introduce small amplitude field modulation. An oscillating magnetic field is superimposed on the DC field by means of small coils, usually built into the cavity walls. When the field is in the vicinity of a resonance line, it is swept back and forth through part of the line, leading to an AC component in the diode current. This AC component is amplified using a frequency selective amplifier tuned to the modulation frequency, thus eliminating a great deal of noise. The modulation amplitude is normally less than the line width. Thus the detected AC signal is proportional to the change in sample absorption as the field is swept. It takes a little practice to get used to looking at first-derivative spectra, but there is a distinct advantage: first-derivative spectra have much better apparent resolution than do absorption spectra. Indeed, second-derivative spectra are even better resolved (though the SNR decreases on further differentiation). The microwave-generating klystron tube requires explanation. A schematic drawing of the klystron is shown in Figure 4.19. There are three electrodes: a heated cathode from which electrons are emitted, an anode to collect the electrons, and a highly negative reflector electrode that sends those electrons which pass through a hole in the anode back to the anode. The motion of the charged electrons from the hole in the anode to the reflector and back to the anode generates an oscillating electric field and thus electromagnetic radiation. The transit time from the hole to the reflector and back again corresponds to the period of oscillation (ν). Thus

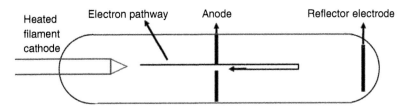

Figure 4.19 Schematic drawing of a microwave-generating klystron tube. *Source:* adapted from [69].

the microwave frequency can be tuned (over a small range) by adjusting the physical distance between the anode and the reflector or by adjusting the reflector voltage. In practice, both methods are used: the metal tube is distorted mechanically to adjust the distance (a coarse frequency adjustment) and the reflector voltage is adjusted as a fine control.

The sample is mounted in the microwave cavity (Figure 4.20). The cavity is a rectangular metal box, exactly one wavelength long. An X-band cavity has dimensions of about $1 \times 2 \times 3$ cm. The electric and magnetic fields of the standing wave are shown in this figure. Note that the sample is mounted in the electric field nodal plane, but at a maximum in the magnetic field. The static field, B, is perpendicular to the sample port. The cavity length is not adjustable, but it must be exactly one wavelength. Thus the spectrometer must be tuned such that the klystron frequency is equal to the cavity resonant frequency. The tune-up procedure usually includes observing the klystron power mode. That is, the frequency is swept across a range that includes the cavity resonance by sweeping the klystron reflector voltage, and the diode detector current is plotted on an oscilloscope or other device. When the klystron frequency is close

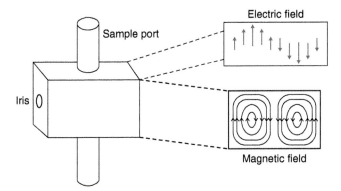

Figure 4.20 Microwave cavity: a rectangular metal box, exactly one wavelength long. *Source:* adapted from [69].

to the cavity resonance, microwave energy is absorbed by the cavity and the power reflected from the cavity to the diode is minimized, resulting in a dip in the power mode. The "cavity dip" is centered on the power mode using the coarse mechanical frequency adjustment, while the reflector voltage is used to fine tune the frequency.

4.4.3 Optimization of Operating Parameters

Determining the ESR spectrum of a sample using a typical CW spectrometer of the sort outlined in Figure 4.18, which is still the most common commercially available type of instrument, involves adjustment of the set of operating parameters described below. In the early days of ESR these adjustments would be carried out using control knobs on a console. Nowadays, of course, the settings are carried out under control of a computer interface. The purpose of these parameters and typical values, however, has remained unchanged. Such spectrometers are quite adequate for studying relatively stable samples. Characterizing transient species by ESR, however, requires substantial modification of commercial instruments or the use of a pulsed spectrometer. There are 12 parameters that must be set or known in recording an ESR spectrum (Table 4.3). Briefly, below, each parameter is discussed and the means used to optimize or measure the parameter described.

4.4.3.1 Microwave Frequency

The resonant microwave frequency reaching the sample is determined by the effective length of the microwave cavity. The actual length is somewhat modified by the influence of the sample tube and the Dewar insert (if controlled temperature operation is required) so that the microwave frequency varies by a few percent even for the same cavity. Since the klystron frequency is locked to the cavity resonant frequency by a suitable feedback circuit, this is not an adjustable parameter. However, to compute a g-value from a spectrum, the frequency must be known quite accurately. This is normally done using a microwave frequency counter installed somewhere in the waveguide circuit.

Table 4.3 Parameters involved in the recording of an ESR spectrum, 12 parameters that must be set or known in recording [69].

Microwave frequency	Center field	Modulation frequency	Modulation phase
Microwave power	Sweep width	First and second harmonic	Signal gain
Sweep time	Field offset	Modulation amplitude	Filter time constant

4.4.3.2 Center Field, Sweep Width, and Field Offset

Once you know, or can guess, the field limits of your spectrum, setting the center field and sweep width values is not very difficult. The center field corresponds to the middle of the spectrum and a sufficiently large sweep width chosen so that all of the spectrum is recorded. If you do not know the field range occupied by your spectrum in advance, the center field must be chosen by educated guess; set the sweep width $2 - 4\times$ greater than the expected width. Hopefully, you will see at least a piece of your spectrum and can make appropriate adjustments to zero in on the correct settings. Most spectrometers measure the magnetic field by a Hall effect probe consisting of a sensor mounted on one of the pole faces of the magnet. However, such estimates of the value of B inside the cavity are not sufficiently accurate to be used for g-factor determinations. There are two ways around this problem: (i) measure the spectrum of a solid free radical such as diphenylpicrylhydrazyl (DPPH), which has a well-known g-value (2.0028), at least once during acquisition of the desired spectrum; or (ii) use of an NMR gauss meter probe inserted in or near the cavity several times during the collection of the spectrum.

4.4.3.3 Sweep Time

In general, the longer the sweep time the better the sensitivity since the filter time constant parameter can be set longer with consequent improvement in SNR. In practice, however, sweep times are usually set in accordance with the expected lifetime of the radical species, the stability of the instrument, and the patience of the operator. Decay of the radical or drift of the spectrometer during a scan is clearly undesirable. The sweep time is most commonly set in the range 4–10 minutes.

4.4.3.4 Modulation Frequency

With most spectrometers, you have a choice of either 100 kHz or a lower frequency of field modulation. The higher frequency generally gives better S/N, but if the lines are unusually sharp (<0.08 G), 100 kHz modulation leads to "side bands," lumps in the line shape that confuse the interpretation of the spectrum. Under such circumstances, use the lower frequency for which the sidebands are closer together and thus less likely to be a problem.

4.4.3.5 Second Harmonic Detection

In most cases, you will use the first harmonic and the normal first-derivative of the absorption spectrum will be presented. If your spectrum has very good S/N and has some regions where you would like better resolution, a second derivative presentation may help. However, second derivatives from second harmonic detection are very costly in terms of S/N ratio and so you really do have to have a strong signal!

4.4.3.6 Modulation Amplitude

Since the absorption signal is usually detected by imposing a 100 kHz field modulation on the static field, the signal disappears when the modulation amplitude is turned to zero. In general, the signal increases more or less in proportion to the modulation amplitude, but eventually the detected lines begin to broaden and then the signal amplitude decays as well. Depending on what you want to optimize, here are some rules: For optimum S/N ratio, but decreased resolution: Modulation amplitude = 2 × line width. For accurate line width measurements: Modulation amplitude = line width/10. For most practical work: Modulation amplitude = line width/3.

4.4.3.7 Modulation Phase

To improve the S/N ratio, the modulation signal is processed by amplification with a tuned amplifier using phase-sensitive detection. This means that the detected signal must not only be at the modulation frequency, but must also be in phase with the modulation. Since the amplifier itself can introduce a bit of phase shift, there is a phase control which, in principle, should be adjusted to maximize the signal amplitude. In practice, this control needs to be adjusted only rarely and in most cases the best approach is to leave it alone.

4.4.3.8 Signal Gain

Adjustment of the signal gain is straightforward. Set the gain sufficiently high that the recorded spectrum is nearly full-scale on the computer displays or recorder. If you have no idea of the strength of your expected signal, a good starting point would be 1×10^4. Too high a gain can result in amplifier overload.

4.4.4 Biological Application of the ESR

The first publication on the use of ESR techniques in connection with biological systems appeared only nine years after the first demonstration of paramagnetic resonant absorption [70]. In that first report, a variety of biological systems, such as leaves, seeds, and tissue preparations, were shown to contain free radicals. A definite correlation was found between the concentration of the radicals and the metabolic activity of the material. This work appears to confirm earlier ideas that free radicals are involved as intermediates in metabolic processes. However, the question as to which free radicals are involved in a given metabolic process proved to be a much more difficult question to answer. Part of the difficulty lies in the nature of the ESR spectra found for biological systems. One must realize that most biological reactions occur within an organized structure such as a membrane. Thus if radicals are produced as intermediates, they will most likely be bound or closely associated with the enzymes which catalyze the specific reaction. These immobilized free radicals give rise to ESR spectra which are similar to those which might be obtained by freezing a solution of an organic radical. Although a strong free radical signal may be seen, little or no hyperfine structure can be resolved.

The immobilization of biological paramagnetic species also presents sensitivity problems, since the "solid-state" spectra generally have much broader lines than the corresponding "solution" spectra. A further difficulty is presented by the fact that almost all biological materials, including *in vivo* samples, contain a high proportion of water. Because of the high dielectric loss of water, there are severe restrictions on sensitivity, sample size, and shape. In spite of the above difficulties, impressive advances involving ESR spectroscopy have been made in certain areas [71].

4.4.4.1 ESR Oximetry

The quantification of oxygen levels in different tissues or in blood is of paramount importance for the diagnosis and treatment of numerous diseases connected with altered supply and/or utilization of oxygen [72]. Besides other methods ESR oximetry has become a widely used technique to determine oxygen levels in tissues, due to its minimal invasiveness, the high accuracy compared to, for example, chemical methods and the possibility to obtain information about the spatial distribution of oxygen levels by using ESR imaging approaches. The theory, instrumentation, and various applications of ESR oximetry have been comprehensively reviewed, for example, by Swartz and Clarkson [73], and by Ahmad and Kuppusamy [74]. In brief, determination of oxygen concentrations or partial oxygen pressures (pO_2) by ESR spectroscopy is based on the interaction of the paramagnetic oxygen molecule with paramagnetic *spin probes* that can be applied either in solid or soluble form. Interaction of the two paramagnetic species, including DD interaction and – being the dominant type of interaction – Heisenberg spin exchange, leads to changes in the ESR line width and the extent of broadening is proportional to the oxygen concentration or pO_2, thus providing means for the accurate quantification of oxygen levels.

Applications of ESR oximetry include, for example, the assessment of tumor oxygenation, reviewed by Gallez et al. [75], monitoring of tissue oxygenation, blood flow and oxidate stress in the brain [76], and quantification of cellular respiration by measuring the oxygen consumption of cells, mitochondria and submitochondrial particles [77]. Although ESR oximetry in general will not be further detailed in this review due to its comprehensive treatment in the literature, one aspect should be stressed here that concerns the use of insoluble, particulate paramagnetic probes for repeated measurements of the oxygenation of specific tissues by ESR spectroscopy. Implantable spin probes for ESR oximetry are, for example, specific types of charcoal, that long-term responsiveness to oxygen in tissues can be further enhanced by coating of the particles [78], or substances like lithium octa-*n*-butoxynaphthalocyanine (LiNc–BuO) microcrystals in the oxygen-permeable and bioinert polymer polydimethylsiloxane

(PDMS) [79]. Further information on the use of particulate materials in ESR oximetry can be found in Khan et al. [80].

4.4.4.2 Direct Detection of Paramagnetic Species

Numerous paramagnetic species playing important roles *in vivo* are relatively stable and therefore directly observable by ESR spectroscopy, like radical species of amino acids in proteins (e.g. the tyrosyl radical), organic radical cofactors in proteins (e.g. semiquinones) and paramagnetic transition metal ions. Some examples for which the reader is referred to the respective literature include the ESR spectroscopy of melanin in dermatology (e.g. in skin tumors) reviewed in [81], the quantification of radiation-induced (accidental as well as from terrorism or war) ESR signals in teeth (ESR dosimetry, reviewed, e.g. in [82]), and other examples where lifetime and concentration of the radical species permit their direct detection like ascorbyl, α-tocopheroxyl, or phenoxyl radicals (see [83]).

Several transition metal ions have been identified to play a crucial role in central nervous system (CNS) diseases like Alzheimer's disease (AD), Parkinson's disease (PD), and prion diseases, what has been reviewed by Kozlowski et al. [84] and Viles [85]. Binding of such metal ions has been implicated in misfolding and/or subsequent amyloid plaque formation of a number of proteins associated with the respective disease. For example, the Cu^{2+} concentration in the neocortical parenchyma of the healthy brain has been shown to be about 70 μM, whereas in amyloid plaques of AD patients concentrations up to 400 μM have been observed, providing strong evidence that Cu^{2+} plays an important role in plaque formation. Consequently, an understanding of the role of such metal ions in CNS diseases, especially concerning the structural features of the ion binding sites and their implications for plaque formation, is of great importance, not least for the development of potential therapeutic drugs. Especially in the case of Cu^{2+}, but also for other paramagnetic transition metal ions, the use of ESR spectroscopy to gain insights into the coordination of the ions within the proteins in their folded and misfolded/aggregated states is straightforward. For example, it has been shown by cw ESR spectroscopy, that Cu^{2+} binds to both the N- and C-terminal parts of mammalian prion proteins and that the number of binding sites – as previously suggested by the use of other techniques – depends on the environmental pH [86]. In this study three pH-dependent binding modes for the C-terminal domain have been identified that have been subjected to a more detailed investigation by the same authors, utilizing more advanced ESR techniques, namely hyperfine sublevel correlation (HYSCORE) spectroscopy and electron nuclear double resonance (ENDOR) spectroscopy, revealing substantial details about the coordination of the copper ions in these binding sites [86]. More examples for investigations on copper binding to prion proteins can be found in a review by Drew and Barnham [87].

4.4.4.3 EPR Revealed the Nitrite Reductase Activity of Myoglobin

Although a cytoprotective effect of low doses of nitrite in the setting of myocardial, liver, kidney, and brain ischemia-reperfusion (I/R) injury had been demonstrated by several research groups, the mechanistic basis of this effect remained a matter of debate for several years. Indications that enzymatic conversion of (exogenous) nitrite to nitric oxide (NO•) by xanthineoxidoreductase is responsible for the protective role of nitrite during such injury could not be corroborated by other groups. Alternative mechanisms how nitrite can be *in vivo* converted to NO• are acidic disproportionation or enzymatic conversion by mitochondrial enzymes, deoxyhemoglobin, and deoxymyoglobin. In a study using myoglobin wild-type (+/+) and knockout (−/−) mice, Hendgen-Cotta et al. [88] demonstrated unambiguously that the nitrite reductase activity of myoglobin is responsible for this effect by regulating respiration and cellular viability in myocardial I/R injury. A key observation in this study was the detection by ESR of a Mb^{15}NO signal – characterized by a doublet hyperfine splitting arising from the ^{15}N nuclear spin of ½ – in myoglobin +/+ hearts after infusion of [^{15}N] nitrite but not of buffer prior to a global ischemia, demonstrating that myoglobin reduces exogenous nitrite to NO•. Furthermore in hearts from myoglobin knockout (−/−) mice no NO-heme complexes could be observed, revealing that myoglobin and not hemoglobin acts as a nitrite reductase in I/R.

4.4.4.4 Mitochondrial Dysfunction in Severe Sepsis

Sepsis, a systemic inflammatory response to infection, is the leading cause of death in critically ill patients caused by multiple organ failure. Although the precise mechanisms underlying this process are still not fully elucidated, bioenergetic failure, in particular mitochondrial dysfunction has been related to sepsis-induced organ failure [89]. Svistunenko et al. [90] addressed this issue in an ESR study on human muscle biopsies taken from critically ill patients with severe sepsis. They used a self-developed spectra deconvolution method to dissect and quantify the contributions from nine different paramagnetic species in low temperature ESR spectra of the tissue samples. The authors found that one of the species identified, a spin-coupled pair of semiquinone radicals (SQ• – SQ•) that concentration seems to be related to the redox state of the mitochondrial complex II, negatively correlates with the illness severity of the patient. Furthermore, the study revealed that a decreased concentration of mitochondrial complex I iron–sulfur centers is linked to mortality. Although the causes of this links remain to be elucidated, this study confirms the close connection between septic shock and mitochondrial dysfunction, in particular complexes I and II of the respiratory chain.

4.4.4.5 Spin Trapping ESR

Free radicals, like nitric oxide, superoxide or hydroxyl radicals, play multiple roles in biological systems, being either involved in signaling processes or representing "unwanted" and potentially harmful side-products of metabolic processes in cells. Although direct detection of such species by ESR would be the ideal case, the exceptionally short lifetime of numerous of these radicals (due to their high reactivity) prevents their direct measurement. This difficulty can be overcome by using the spin trapping methodology.

Reactive oxygen species (ROS) play important roles in cell signaling and homeostasis and are natural by-products of the normal oxygen metabolism. Nevertheless, unfavorable conditions, e.g. environmental stress, can lead to dramatically increased ROS levels that – due to the high reactivity of these molecules – may result in significant damage to the cell. Increased ROS levels are often caused by diseases, for example, in inflammatory responses including cardiovascular disease and ischemic injury including stroke and heart attack. Furthermore, ROS are according to the free-radical theory thought to be a major contributor to the organism's functional decline in the course of aging, rendering their investigation to be of utmost importance. The use of ESR spectroscopy in combination with the spin trapping approach for the detection and quantification of ROS and, more general for evaluating the redox state *in vivo* has been extensively reviewed by Swartz et al. [91]. In the following section, three examples are given that highlight different aspects of spin trapping of ROS *in vivo*.

Lundqvist et al. [92] investigated the amount of superoxide anion radicals produced by human neutrophils interacting with viable Staphylococcus bacteria using the spin traps 5,5-dimethyl-1-pyrroline-*N*-oxide (DMPO) and DEPMPO. They incubated isolated human granulocytes with viable *Staphylococcus aureus* and *Staphylococcus epidermidis* bacteria in the presence of the DMPO or DEPMPO and found significantly increased levels of spin adducts in the presence of bacteria. Consideration of the spin adduct life times and careful interpretation of the ESR spectra taken after different incubation times furthermore allowed to identify the superoxide radical ($O_2^{\cdot-}$) being the primary radical product from neutrophils.

In a spin trapping study using phenylbutylnitrone (PBN), Vrbjar et al. [93] investigated radical-induced lipid peroxidation caused by postischemic reperfusion in isolated rat hearts. They were able to demonstrate formation and release of alkoxy radicals within the first 15 minutes of reperfusion following 30 minutes of ischemia. Furthermore, they could observe that decline of the radical (after 10 minutes of reperfusion) was accompanied by recovery of heart function and that the presence of PBN led to improved functional recovery compared to untreated (PBN-free) controls. In addition, reduced radical concentrations correlated with increased improvement of heart functional recovery were observed when the radical scavengers mercaptopropionylglycine (MPG) or vitamin E were applied.

References

1 Chizhik, V.I., Chernyshev, Y.S., Donets, A.V. et al. (2014). *Magnetic Resonance and Its Applications*. Springer.

2 Arkad'ev, V.K. (1913). The reflection of electric waves from a wire. *Soviet Physics – JETP* 45A (45): 312.

3 Cleeton, C.E. and Williams, N.H. (1934). Electromagnetic waves of 1.1 cm wavelength and the absorption spectrum of ammonia. *Physical Review* 45 (4): 234.

4 Rabi, I.I., Millman, S., Kusch, P., and Zacharias, J.R. (1939). The molecular beam resonance method for measuring nuclear magnetic moments. The magnetic moments of $_3Li^6$, $_3Li^7$ and $_9F^{19}$. *Physical Review* 55 (6): 526.

5 Bloch, F., Hansen, W.W., and Martin, E. (1946). Packard nuclear induction. *Physical Review* 127.

6 Purcell, E.M., Torrey, H.C., and Pound, R.V. (1946). Resonance absorption by nuclear magnetic moments in a solid. *Physical Review* 69: 37–38.

7 Dehmelt, H. and Krueger, H. (1950). Kernquadrupolfrequenzen in festem Dichloraethylen. *Naturwissenschaften* 37: 111–112.

8 Lei, K., Heidari, H., Mak, P.I. et al. (2016). A handheld 50pM-sensitivity micro-NMR CMOS platform with B-field stabilization for multi-type biological/chemical assays. In: *IEEE International Solid-State Circuits Conference (ISSCC)*, 474–475. IEEE.

9 Lei, K.-M., Heidari, H., Mak, P.-I. et al. (2016). A handheld high-sensitivity micro-NMR CMOS platform with B-field stabilization for multi-type biological/chemical assays. *IEEE Journal of Solid-State Circuits* 52 (1): 284–297.

10 Yin, Z., Bonizzoni, E., and Heidari, H. (2018). Magnetoresistive biosensors for on-chip detection and localization of paramagnetic particles. *IEEE Journal of Electromagnetics, RF and Microwaves in Medicine and Biology* 2 (3): 179–185.

11 Shah, S. and Heidari, H. (2017). On-chip magnetoresistive sensors for detection and localization of paramagnetic particles. In: *2017 IEEE SENSORS*, 1–3. IEEE.

12 Zuo, S., Nazarpour, K., and Heidari, H. (2018). Device modeling of MgO-barrier tunneling magnetoresistors for hybrid spintronic-CMOS. *IEEE Electron Device Letters* 39 (11): 1784–1787.

13 Gordon, J.P., Zeiger, H.J., and Townes, C.H. (1955). The maser – new type of microwave amplifier, frequency standard, and spectrometer. *Physical Review* 99: 1264.

14 Basov, N.G. and Prokhorov, A.M. (1955). Molecular generator and amplifier. *Physics-Uspekhi (Advances in Physical Sciences)* 57 (3): 481.

15 Bloembergen, N. (1956). Proposal for a new type solid state maser. *Physical Review* 104 (2): 324.

16 Schawlow, A.L. and Townes, C.H. (1958). Infrared and optical masers. *Physical Review* 112: 1940.

17 Maiman, T.H. (1960). Stimulated optical radiation in ruby. *Nature* 187: 493–494.

18 Javan, A., Bennett, W.R. Jr., and Herriott, D.R. (1961). Population inversion and continuous optical maser oscillation in a gas discharge containing a He–Ne mixture. *Physical Review Letters* 6 (3): 106.

19 Chary, K.V.R. and Govil, G. (2008). *NMR in Biological Systems: From Molecules to Human.* Springer Science & Business Media.

20 Gerlach, W. and Stern, O. (1922). Das magnetische Moment des Silberatoms. *Zeitschrift für Physik* 9: 353–355.

21 Estermann, I., Frisch, R., and Stern, O. (1933). Magnetic moment of the proton. *Nature* 132 (3326): 169.

22 Constantinides, C. (2014). *Magnetic Resonance Imaging: The Basics.* CRC Press.

23 Suryan, G. (1951). Nuclear resonance in flowing liquids. *Proceedings of the Indian Academy of Sciences – Section A* 33: 107.

24 Kuperman, V. (2000). *Magnetic Resonance Imaging: Physical Principles and Applications.* Elsevier.

25 Singer, J.R. (1959). Blood flow rates by nuclear magnetic resonance measurements. *Science* 130 (3389): 1652–1653.

26 Singer, J.R. (1960). Flow rates using nuclear or electron paramagnetic resonance techniques with applications to biological and chemical processes. *Journal of Applied Physics* 31 (1): 125–127.

27 Bowman, R.L. and Kudravcev, V. (1959). Blood flowmeter utilizing nuclear magnetic resonance. *IRE Transactions on Medical Electronics* ME-6 (4): 267–269.

28 Battocletti, J.H. (1986). Blood flow measurement by NMR. *Critical Reviews in Biomedical Engineering* 13 (4): 311–367.

29 Young, I.R., Burl, M., Clarke, G.J. et al. (1981). Magnetic resonance properties of hydrogen: imaging the posterior fossa. *American Journal of Roentgenology* 137 (5): 895–901.

30 Crooks, L., Sheldon, P., Kaufman, L. et al. (1982). Quantification of obstructions in vessels by nuclear magnetic resonance (NMR). *IEEE Transactions on Nuclear Science* 29 (3): 1181–1185.

31 Gullberg, G.T., Wehrli, F.W., Shimakawa, A., and Simons, M.A. (1987). MR vascular imaging with a fast gradient refocusing pulse sequence and reformatted images from transaxial sections. *Radiology* 165 (1): 241–246.

32 Keller, P.J., Drayer, B.P., Fram, E.K. et al. (1989). MR angiography with two-dimensional acquisition and three-dimensional display. Work in progress. *Radiology* 173 (2): 527–532.

33 Laub, G.A. and Kaiser, W.A. (1988). MR angiography with gradient motion refocusing. *Journal of Computer Assisted Tomography* 12 (3): 377–382.

34 Dumoulin, C.L., Cline, H.E., Souza, S.P. et al. (1989). Three-dimensional time-of-flight magnetic resonance angiography using spin saturation. *Magnetic Resonance in Medicine* 11 (1): 35–46.

35 Ruggieri, P.M., Laub, G.A., Masaryk, T.J., and Modic, M.T. (1989). Intracranial circulation: pulse-sequence considerations in three-dimensional (volume) MR angiography. *Radiology* 171 (3): 785–791.

36 Nägele, T., Klose, U., Grodd, W. et al. (1994). The effects of linearly increasing flip angles on 3D inflow MR angiography. *Magnetic Resonance in Medicine* 31 (5): 561–566.

37 Parker, D.L., Yuan, C., and Blatter, D.D. (1991). MR angiography by multiple thin slab 3D acquisition. *Magnetic Resonance in Medicine* 17 (2): 434–451.

38 Edelman, R.R., Mattle, H.P., Wallner, B. et al. (1990). Extracranial carotid arteries: evaluation with "black blood" MR angiography. *Radiology* 177 (1): 45–50.

39 Nishimura, D.G., Macovski, A., Pauly, J.M., and Conolly, S.M. (1987). MR angiography by selective inversion recovery. *Magnetic Resonance in Medicine* 4 (2): 193–202.

40 Edelman, R.R., Chien, D., and Kim, D. (1991). Fast selective black blood MR imaging. *Radiology* 181 (3): 655–660.

41 Zerhouni, E.A., Parish, D.M., Rogers, W.J. et al. (1988). Human heart: tagging with MR imaging – a method for noninvasive assessment of myocardial motion. *Radiology* 169 (1): 59–63.

42 Axel, L. and Dougherty, L. (1989). MR imaging of motion with spatial modulation of magnetization. *Radiology* 171 (3): 841–845.

43 Mosher, T.J. and Smith, M.B. (1990). A DANTE tagging sequence for the evaluation of translational sample motion. *Magnetic Resonance in Medicine* 15 (2): 334–339.

44 Kuperman, V.Y., Ehrichs, E.E., Jaeger, H.M., and Karczmar, G.S. (1995). A new technique for differentiating between diffusion and flow in granular media using magnetic resonance imaging. *Review of Scientific Instruments* 66 (8): 4350–4355.

45 Moran, P.R. (1982). A flow velocity zeugmatographic interlace for NMR imaging in humans. *Magnetic Resonance Imaging* 1 (4): 197–203.

46 Dumoulin, C.L. and Hart, H.R. Jr. (1986). Magnetic resonance angiography. *Radiology* 161 (3): 717–720.

47 Prince, M.R. (1994). Gadolinium-enhanced MR aortography. *Radiology* 191 (1): 155–164.

48 Adamis, M.K., Li, W., Wielopolski, P.A. et al. (1995). Dynamic contrast-enhanced subtraction MR angiography of the lower extremities: initial evaluation with a multisection two-dimensional time-of-flight sequence. *Radiology* 196 (3): 689–695.

49 Alley, M.T., Shifrin, R.Y., Pelc, N.J., and Herfkens, R.J. (1998). Ultrafast contrast-enhanced three-dimensional MR angiography: state of the art. *Radiographics* 18 (2): 273–285.

50 Earls, J.P., Rofsky, N.M., DeCorato, D.R. et al. (1996). Breath-hold single-dose gadolinium-enhanced three-dimensional MR aortography: usefulness of a timing examination and MR power injector. *Radiology* 201 (3): 705–710.

51 Wilman, A.H., Riederer, S.J., King, B.F. et al. (1997). Fluoroscopically triggered contrast-enhanced three-dimensional MR angiography with elliptical centric view order: application to the renal arteries. *Radiology* 205 (1): 137–146.

52 Foo, T.K., Saranathan, M., Prince, M.R., and Chenevert, T.L. (1997). Automated detection of bolus arrival and initiation of data acquisition in fast, three-dimensional, gadolinium-enhanced MR angiography. *Radiology* 203 (1): 275–280.

53 Zavoisky, E. (1945). Spin-magnetic resonance in paramagnetics. *Journal of Physics USSR* 9: 211–245.

54 Bleaney, B. and Stevens, K.W.H. (1953). Paramagnetic resonance. *Reports on Progress in Physics* 16 (1): 108.

55 Bowers, K.D. and Owen, J. (1955). Paramagnetic resonance II. *Reports on Progress in Physics* 18 (1): 304.

56 Bagguley, D.M.S. and Owen, J. (1957). Microwave properties of solids. *Reports on Progress in Physics* 20 (1): 304.

57 Abragam, A. and Bleaney, B. (1970). *Electron Paramagnetic Resonance of Transition Ions*. Oxford University Press.

58 Abragam, A. (1961). *The Principles of Nuclear Magnetism*. Oxford University Press.

59 Ingram, D.J.E. (1958). *Free Radicals as Studied by Electron Spin Resonance*. Academic Press.

60 Pake, G.E. (1962). *Paramagnetic Resonance: An Introductory Monograph*. WA Benjamin.

61 Baird, J.C. and Bersohn, M. (1966). *An Introduction to Electron Paramagnetic Resonance*. Benjamin Inc.

62 Ayscough, P.B. (1967). *Electron Spin Resonance in Chemistry*, 438. London: Methuen.

63 Assenheim, H.M. (2014). *Introduction to Electron Spin Resonance*. Springer.

64 Alger, R.S. (1968). *Electron Paramagnetic Resonance: Techniques and Applications*. Interscience.

65 Weil, J.A. and Bolton, J.R. (2007). *Electron Paramagnetic Resonance: Elementary Theory and Practical Applications*. Wiley.

66 Atherton, N.M. (1993). *Principles of Electron Spin Resonance*. Ellis Horwood Limited.

67 Gordy, W. (1980). *Theory and Applications of Electron Spin Resonance*. Wiley.

68 Eaton, G.R. and Eaton, S.S. (1998). *Foundations of Modern EPR*. World Scientific.

69 Rieger, P. (2007). *Electron Spin Resonance: Analysis and Interpretation*. Royal Society of Chemistry.

70 Commoner, B., Townsend, J., and Pake, G.E. (1954). Free radicals in biological materials. *Nature* 174 (4432): 689.

71 Swartz, H.M., Bolton, J.R., and Borg, D.C. (1972). *Biological Applications of Electron Spin Resonance*. Wiley-Interscience.

72 Klare, J.P. (2012). Biomedical applications of electron paramagnetic resonance (EPR) spectroscopy. *Biomedical Spectroscopy and Imaging* 1 (2): 101–124.

73 Swartz, H.M. and Clarkson, R.B. (1998). The measurement of oxygen in vivo using EPR techniques. *Physics in Medicine & Biology* 43 (7): 1957.

74 Ahmad, R. and Kuppusamy, P. (2010). Theory, instrumentation, and applications of electron paramagnetic resonance oximetry. *Chemical Reviews* 110 (5): 3212–3236.

75 Gallez, B., Baudelet, C., and Jordan, B.F. (2004). Assessment of tumor oxygenation by electron paramagnetic resonance: principles and applications. *NMR in Biomedicine* 17 (5): 240–262.

76 Liu, S., Timmins, G.S., Shi, H. et al. (2004). Application of in vivo EPR in brain research: monitoring tissue oxygenation, blood flow, and oxidative stress. *NMR in Biomedicine* 17 (5): 327–334.

77 Presley, T., Kuppusamy, P., Zweier, J.L., and Ilangovan, G. (2006). Electron paramagnetic resonance oximetry as a quantitative method to measure cellular respiration: a consideration of oxygen diffusion interference. *Biophysical Journal* 91 (12): 4623–4631.

78 Gallez, B., Jordan, B.F., and Baudelet, C. (1999). Microencapsulation of paramagnetic particles by pyrroxylin to preserve their responsiveness to oxygen when used as sensors for in vivo EPR oximetry. *Magnetic Resonance in Medicine: An Official Journal of the International Society for Magnetic Resonance in Medicine* 42 (1): 193–196.

79 Meenakshisundaram, G., Eteshola, E., Pandian, R.P. et al. (2009). Oxygen sensitivity and biocompatibility of an implantable paramagnetic probe for repeated measurements of tissue oxygenation. *Biomedical Microdevices* 11 (4): 817–826.

80 Khan, N., Williams, B.B., Hou, H. et al. (2007). Repetitive tissue pO_2 measurements by electron paramagnetic resonance oximetry: current status and future potential for experimental and clinical studies. *Antioxidants & Redox Signaling* 9 (8): 1169–1182.

81 Plonka, P.M. (2009). Electron paramagnetic resonance as a unique tool for skin and hair research. *Experimental Dermatology* 18 (5): 472–484.

82 Swartz, H.M., Khan, N., Buckey, J. et al. (2004). Clinical applications of EPR: overview and perspectives. *NMR in Biomedicine* 17 (5): 335–351.

83 Brustolon, M. (2009). *Electron Paramagnetic Resonance: A Practitioner's Toolkit*. Wiley.

84 Kozlowski, H., Luczkowski, M., Remelli, M., and Valensin, D. (2012). Copper, zinc and iron in neurodegenerative diseases (Alzheimer's, Parkinson's and prion diseases). *Coordination Chemistry Reviews* 256 (19–20): 2129–2141.

85 Viles, J.H. (2012). Metal ions and amyloid fiber formation in neurodegenerative diseases. Copper, zinc and iron in Alzheimer's, Parkinson's and prion diseases. *Coordination Chemistry Reviews* 256 (19–20): 2271–2284.

86 Cereghetti, G.M., Schweiger, A., Glockshuber, R., and Van Doorslaer, S. (2001). Electron paramagnetic resonance evidence for binding of Cu^{2+} to the C-terminal domain of the murine prion protein. *Biophysical Journal* 81 (1): 516–525.

87 Drew, S.C. and Barnham, K.J. (2008). Biophysical investigations of the prion protein using electron paramagnetic resonance. In: *Prion Protein Protocols*, 173–196. Springer.

88 Hendgen-Cotta, U.B., Merx, M.W., Shiva, S. et al. (2008). Nitrite reductase activity of myoglobin regulates respiration and cellular viability in myocardial ischemia-reperfusion injury. *Proceedings of the National Academy of Sciences* 105 (29): 10256–10261.

89 Brealey, D., Brand, M., Hargreaves, I. et al. (2002). Association between mitochondrial dysfunction and severity and outcome of septic shock. *The Lancet* 360 (9328): 219–223.

90 Svistunenko, D.A., Davies, N., Brealey, D. et al. (2006). Mitochondrial dysfunction in patients with severe sepsis: an EPR interrogation of individual respiratory chain components. *Biochimica et Biophysica Acta (BBA) – Bioenergetics* 1757 (4): 262–272.

91 Swartz, H.M., Khan, N., and Khramtsov, V.V. (2007). Use of electron paramagnetic resonance spectroscopy to evaluate the redox state in vivo. *Antioxidants & Redox Signaling* 9 (10): 1757–1772.

92 Lundqvist, H., Dånmark, S., Johansson, U. et al. (2008). Evaluation of electron spin resonance for studies of superoxide anion production by human neutrophils interacting with *Staphylococcus aureus* and *Staphylococcus epidermidis*. *Journal of Biochemical and Biophysical Methods* 70 (6): 1059–1065.

93 Vrbjar, N., Zöllner, S., Haseloff, R.F. et al. (1998). PBN spin trapping of free radicals in the reperfusion-injured heart. Limitations for pharmacological investigations. In: *Myocardial Ischemia and Reperfusion*, 107–115. Springer.

5

SQUID Sensors

5.1 Introduction

Superconducting quantum interference devices (SQUIDs) have been a key factor in the development and commercialization of ultrasensitive electric and magnetic measurement systems. In many cases, SQUID instrumentation offers the ability to make measurements where no other methodology is possible. In addition to measuring magnetic fields, SQUID sensors can be configured to measure a wide variety of electromagnetic properties. There are many ways to measure magnetic fields and properties [1–6]. Sensing methods have been based on the use of induction coils, flux gate magnetometers, magneto-resistive and Hall effect magnetometers, magneto-optical magnetometers, and optically pumped magnetometers. Sensitivities range from microtesla to picotesla levels.

5.1.1 History

In 1820, Hans Oerstedt found and wrote down that a current flowing through a volume produces a magnetic field. Such currents also occur in the human body producing biomagnetic fields detectable outside the body [7]. One source of weak fluctuating fields are the small ion currents in living materials. These currents are produced by, for example, large masses of excitable, synchronously firing tissue, such as heart muscles. Baule and McFee [8] first detected the MCG in 1962. A more sensitive system was built at MIT, consisting of a shielded room [9] and a SQUID magnetometer [9, 10]. The Low Temperature Laboratory at the Helsinki University of Technology was pioneering the development of biomagnetic multichannel instrumentation. The first 4-channel SQUID gradiometer was built in 1983 [11], a 7-channel gradiometer in 1986 [12], and a 24-channel planar gradiometer in 1989 [13, 14]. All these systems were designed for and used in the Otaniemi magnetically shielded room (MSR)

Magnetic Sensors for Biomedical Applications, First Edition. Hadi Heidari and Vahid Nabaei.
© 2020 by The Institute of Electrical and Electronics Engineers, Inc.
Published 2020 by John Wiley & Sons, Inc.

[15]. Based on this experience and knowledge, the company Neuromag Ltd. was established in Helsinki in 1989, which is nowadays with Elekta Neuromag. The multichannel device KRENIKON (Siemens AG, Erlangen, Germany) with 31-channels ran in 1988 [16] but the commercial version with 37 channels ran in 1989 [17]. A double-dewar MEG-system with altogether 28 simultaneously operating SQUID-channels with noise compensation and software gradiometers was developed by the Dornier company in 1990 [18, 19].

An overview of the principles and design of magnetic sensors and magnetometers is given in [20]. The most sensitive magnetic flux detector is the SQUID. This device, operating at cryogenic temperatures with quantum-limited sensitivity, has demonstrated field resolution at the 10^{-17} T level [21]. A large number of applications configure the SQUID as a magnetometer. However, their sensitivity requires proper attention to cryogenics and environmental noise. This applies to not just laboratory applications, but to every potential use of SQUID sensor. The noise can be described as the sum of frequency-independent (white) and frequency-dependent ($1/f$) terms. There are many exotic uses for SQUID sensors. The ability of a SQUID sensor to measure changes in magnetic fields and currents is based on four effects: Superconductivity, Meissner effect, Flux quantization, and Josephson effect. Typically, a SQUID is a ring of superconductors interrupted by one or more Josephson junctions.

SQUID combines the physical phenomena of flux quantization and Josephson tunneling. First predicted by F. London [22], flux quantization was observed experimentally by Deaver and Fairbank [23] and Doll and Näbauer [24] in 1961. They showed that the flux contained in a closed superconducting loop is quantized in units of the flux quantum $\Phi_0 = h/2e \approx 2.07 \times 10^{-15}$ Wb. Here, $h \equiv 2\pi\hbar$ is Planck's constant, and e is the electronic charge [25]. Superconductivity represents a thermodynamic state – existing below a critical temperature T_C – in which, e.g. current is carried by pairs of electrons with opposite momentum and spin, so-called Cooper pairs [26]. For metallic low-temperature superconductors (LTS), like the most widely used Nb, T_C is usually below 10 K. Low operation temperatures permit very sensitive measurements but require the use of cryogenics, although high-temperature superconductors (HTS) have relaxed demands on the cooling system.

5.2 SQUID Fundamentals

5.2.1 Josephson Junctions

The current in a superconductor is carried by so-called Cooper pairs. Since these pairs have zero spin, they follow boson statistics [26]. As a consequence, they all condense in the same quantum state and can be described by a collective superconducting wave function $\Psi = \Psi_0 \cdot \exp(i\phi)$, with $\phi(x, t)$ being the time- and space-dependent phase and $n_s = |\Psi|^2$ the Cooper pair density. If two superconductors are weakly connected, Cooper pairs can exchange between them. There are different types of how these weak links or junctions can be arranged. Probably the most important type is the so-called SIS Josephson tunnel junction, where a thin insulating barrier (I) is placed between two superconductors (S). The current through the Josephson junction is described by the first Josephson equation $I_C = I_{C,0}\sin(\varphi)$, with $\varphi = \phi_1 - \phi_2$ being the phase difference across the junction [27]. Here $I_{C,0}$ is the junction's maximum critical current which is determined by the thickness of the insulating barrier t_{ox}, the junction area A_{JJ}, and the operation temperature T.

When the maximum critical current is exceeded, the phase difference across the junction will evolve over time and a DC voltage across the junction appears. It is described by the second Josephson equation [27]

$$\frac{\partial \varphi}{\partial t} = \frac{2e}{\hbar} \cdot V_{DC} = \frac{2\pi}{\Phi_0} \cdot V_{DC} \tag{5.1}$$

where $\hbar = 1.055 \times 10^{-34}$ Js is the reduced Planck's constant. Please note that subsequently V represents the time averaged DC voltage over the junction. In fact the Josephson current oscillates with the Josephson frequency $2\pi V_{DC}/\Phi_0$, when biased with $I > I_{C,0}$.

A typical current–voltage characteristic of an undamped Josephson junction exhibits a hysteresis, as shown in Figure 5.1 (left). A measure for this hysteresis is the McCumber parameter [28, 29]

$$\beta_C = \frac{2\pi I_C R^2 C_{JJ}}{\Phi_0} \tag{5.2}$$

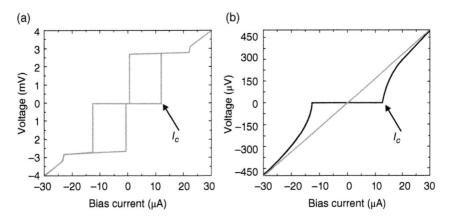

Figure 5.1 (a) Current–voltage characteristics of an undamped and (b) of a damped (shunted) Josephson tunnel junction. The critical current of the junction I_C is about 10 µA, as indicated. For large bias currents, the characteristic of the shunted junction converges into an ohmic behavior, given by the shunt resistor value. Source: adapted from [26].

In order to avoid the hysteresis and therefore to obtain a single valued characteristic depicted in the right panel of Figure 5.1, an additional shunt resistor R_S is usually placed across the junction to damp its dynamics, which is fulfilled for the condition $\beta_C < 1$. The dynamics of Josephson junctions are typically described in the so-called RCSJ (resistively and capacitively shunted junction) model. Therein a real Josephson junction is composed of an ideal one with additional resistance R and capacitance C_{JJ} in parallel, describing the tunneling of normal electrons in the voltage state and the displacement current over the capacitance between the two superconducting electrodes, respectively. Due to finite thermal energy at temperatures $T > 0$, the I–V characteristic of a non-hysteretic junction is noise-rounded for currents of about I_C, as can be seen in Figure 5.1 (right). The ratio between thermal energy $k_B T$ and Josephson coupling energy $E_J = I_C \Phi_0 / 2\pi$ describes the strength of noise-rounding due to thermal noise of the shunt resistor [30, 31] and is known as the noise parameter

$$\Gamma = \frac{k_B T}{E_J} = \frac{2\pi k_B T}{I_C \Phi_0} \tag{5.3}$$

Here, k_B is Boltzmann's constant. In LTS DC SQUIDs the influence due to thermal noise-rounding is typically neglected for $\Gamma < 0.05$.

5.2.2 DC SQUIDs

The DC SQUID (Figure 5.2a) consists of two Josephson junctions connected in parallel on a superconducting loop. In most – but not all – operating schemes, each junction has a resistive shunt to eliminate hysteresis on its I–V characteristic, and we begin with a brief review of the RCSJ. In the absence of any added

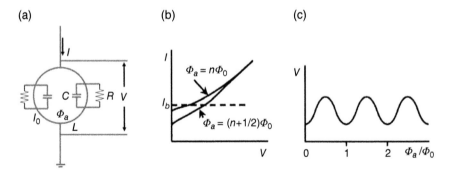

Figure 5.2 The DC SQUID: (a) Schematic; (b) current–voltage characteristics at integer and half-integer values of applied flux; the operation point is set by the bias current I_b; (c) voltage vs. flux Φ_a/Φ_0 for constant bias current. *Source:* adapted from [25].

damping resistance, generally speaking, the I–V characteristic of a Josephson junction is hysteretic. As the current is raised from zero, eventually the voltage switches to $2\Delta/e$; the voltage remains there as the current is reduced until it attains a relatively low value, at which the voltage switches back to zero. For a junction with critical current I_0 and self-capacitance C, this hysteresis is removed by the addition of a shunt of resistance R provided that [28, 29]

$$\beta_C = \frac{2\pi I_0 R^2 C}{\Phi_0} = \omega_J RC \le 1 \tag{5.4}$$

Here, $\omega_J/2\pi \equiv I_0 R/\Phi_0$ is the Josephson frequency at voltage $I_0 R$. For $\beta_C \ll 1$, the I–V characteristic is given by

$$V = R\left(I^2 - I_0^2\right)^{1/2} \tag{5.5}$$

which tends asymptotically to IR for $I \gg I_0$. The resistive shunt has an associated Nyquist noise current with a spectral density $S_I(f) = 4k_B T/R$, where k_B is Boltzmann's constant. This noise has two effects. First, it rounds the I–V characteristic at low voltages and reduces the apparent critical current. To maintain a reasonable degree of Josephson coupling, one requires $I_0\Phi_0/2\pi \gg k_B T$; $I_0\Phi_0/2\pi$ is the Josephson coupling energy. This inequality is conveniently written as $\Gamma \equiv 2\pi k_B T/I_0\Phi_0 \ll 1$. Second, the noise current induces a voltage noise across the junction at nonzero voltages. Computer simulations of the DC SQUID are invariably based on the RCSJ. When the DC SQUID is biased at an appropriate constant current I_b (Figure 5.2b), changes in applied magnetic flux cause the voltage to swing between two extrema, producing the oscillations with period Φ_0 shown in Figure 5.2c.

The maximum response to a small flux change $\delta\Phi_a \ll \Phi_0$ is obtained when $\Phi_a \approx (2n + 1)\Phi_0/4$, where the flux-to-voltage transfer coefficient $V_\Phi \equiv |(\partial V/\partial\Phi_a)_I|$ is a maximum. The resulting voltage change $\delta V = V_\Phi\, \delta\Phi_a$ is approximately linear in $\delta\Phi_a$ in this regime. Detailed computer simulations [32] show that the response is optimized when β_C is at a value just below the onset of hysteresis and when

$$\beta_L = \frac{2LI_0}{\Phi_0} = 1 \tag{5.6}$$

where L is the loop inductance and I_0 is the critical current per junction. For this value of β_L the critical current at $(2n + 1)\,\Phi_0/2$ is approximately one-half that at $n\Phi_0$. The transfer function is predicted to be

$$V_\Phi \approx \frac{R}{L} \approx \frac{1}{(\pi LC)^{1/2}} \tag{5.7}$$

where $\beta_c = \beta_L = 1$.

Nyquist noise in the shunt resistors imposes an upper limit on the SQUID inductance, namely $\Phi_0^2/2L \gg 2\pi k_B T$. Quantum interference is unobservable unless this criterion is satisfied. In addition, the Nyquist noise introduces a white voltage noise across the SQUID with a flux- and bias-current-dependent spectral density $S_v(f)$, which is equivalent to a flux noise with spectral density $S_\Phi(f)$ = $S_v(f)/V_\Phi^2$. Under optimum conditions, one finds [32]

$$S_\Phi(f) \approx \frac{16 k_B T L^2}{R} \tag{5.8}$$

It is often useful to characterize SQUIDs in terms of their noise energy

$$\varepsilon(f) = \frac{S_{\Phi(f)}}{2L}, \text{whence} \tag{5.9}$$

$$\varepsilon(f) \approx \frac{9 k_B T L}{R} \approx 16 k_B T (LC)^{1/2} \tag{5.10}$$

It should be noted, $\beta_c = \beta_L = 1$ to obtain the last expression. The white noise of a large number of low-T_C SQUIDs has been found to be generally in good agreement with these predictions. The $\varepsilon(f)$ is not a complete characterization of the white noise in the DC SQUID since it does not take into account the accompanying noise current in the SQUID loop, which induces a noise voltage into any input circuit coupled to the SQUID. Low-T_C DC SQUIDs are almost invariably fabricated from thin, polycrystalline films of Nb, most often with Nb-AlO$_x$-Nb junctions fabricated in situ, that is, by sequential deposition and Al oxidation without breaking vacuum. This process can yield spreads in critical current of only a few percent across the wafer. The trilayers are patterned with reactive ion etching, and their sides protected by anodization. Subsequently, thin-film shunt resistors of Pd, Mo, W, or AuCu are deposited. The multiturn, spiral input coil – to be connected to an appropriate input circuit – is integrated on top of the SQUID washer, insulated from it by silicon oxide. The washer acts as a ground plane for the coil, providing efficient flux coupling into the SQUID. A second coil to provide flux modulation and feedback may also be fabricated on the chip.

For high-T_C DC SQUIDs in the low-fluctuation regime $\Gamma \ll 1$, the $V_\Phi \approx R/L$ and $\varepsilon(f) \approx 9 k_B T L/R$ remain valid. However, since high-T_C Josephson junctions are self-shunted and require no external shunt, the extensions of these two expressions differ from those for the low-T_C case. We assume that $I_0 R = V_c$ for a given technology, and that this expression remains valid as the width of the junction (parallel to the barrier or weak-link interlayer) is varied. Thus, one can vary I_0 simply by changing the width. Under these circumstances, with $\beta_L = 2LI_0/\Phi_0 = 1$, it has been found [25]

$$V_\Phi \approx \frac{R}{L} \approx \frac{2V_c}{\Phi_0} \tag{5.11}$$

and

$$\varepsilon(f) \approx \frac{4k_B T \Phi_0}{V_c}.$$ (5.12)

Remarkably, for fixed V_0 and in the limit $\Gamma \ll 1$, V_Φ and hence $\varepsilon(f)$ are independent of the DC SQUID parameters. Thus, V_0 becomes an alternative figure of merit for the high-T_C DC SQUID. High-T_C DC SQUIDs are fabricated from thin films of YBa$_2$Cu$_3$O$_{7-x}$ (YBCO), mostly on bicrystals to form grain boundary junctions. Devices with integrated input coils following the design for low-T_C SQUIDs have been successfully made and tested, but their production is nowhere near as routine as that of Nb-based devices. Thus, single-layer devices are more commonly used. Except for high-frequency applications – above a few megahertz – the DC SQUID is almost invariably operated in a flux-locked feedback loop. A detailed account appears in [25]. Feedback serves several purposes: it linearizes the response of the SQUID to applied flux, it enables one to track changes in flux corresponding to many flux quanta, and it enables one to detect changes in flux corresponding to a tiny fraction of a flux quantum. An example – in which a modulating flux is applied to the SQUID at frequencies f_m ranging from perhaps 100 kHz to 10 MHz – is shown in Figure 5.3.

The flux has a peak-to-peak value of $\Phi_0/2$. When the flux in the SQUID is $n\Phi_0$, the resulting voltage is a "rectified" sine wave, as shown in Figure 5.4a. When this voltage is connected to a lock-in detector referenced to f_m, the output is zero. On the other hand, when the flux is $(n + 1/4)\Phi_0$ (Figure 5.4b), the output from the lock-in is a maximum. Thus, as one increases the flux from $n\Phi_0$ to $(n + 1/4)\Phi_0$, the output from the lock-in steadily increases; if instead we decrease the flux from $n\Phi_0$ to $(n - 1/4)\Phi_0$, the output from the lock-in is negative (Figure 5.4c).

After integration, the signal from the lock-in is coupled, via a resistor, to the same coil as that producing the flux modulation. A flux $\delta\Phi_a$ applied to the SQUID results in an opposing flux $-\delta\Phi_a$ from the feedback loop to maintain

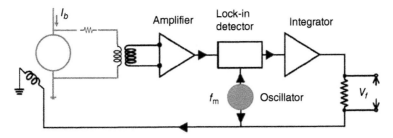

Figure 5.3 Flux modulation and feedback circuit for the DC SQUID. The modulation oscillator shown operates at the frequency f_m. *Source:* adapted from [25].

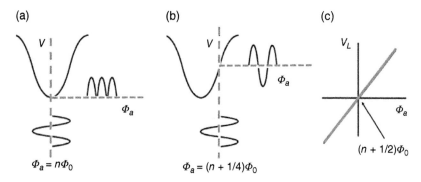

Figure 5.4 Flux modulation scheme showing voltage across the DC SQUID for (a) $\Phi_a = n\Phi_0$ and (b) $\Phi_a = (n + 1/4)\Phi_0$. The output V_L from the lock-in detector versus Φ_a is shown in (c). *Source:* adapted from [25].

(ideally) a constant flux in the SQUID, while producing an output voltage across the resistor proportional to $\delta\Phi_a$. With judicious design, the electronics adds very little noise to the intrinsic noise of a DC SQUID at 4.2 K. The oscillating voltage across the SQUID is amplified by either a tank circuit or more commonly a transformer that boosts the dynamic resistance of the SQUID – typically of the order of 10 Ω – to the value required to optimize the noise temperature of the preamplifier which is typically a few kilohms. Since the noise temperature of the optimally coupled preamplifier is typically about 1 K at frequencies of a few hundred kilohertz, it does not contribute significantly to the system noise. Typically, the white flux noise for a low-T_C SQUID at 4.2 K is $10^{-6}\Phi_0$ Hz$^{-1/2}$ with an onset of $1/f$ noise below 1 Hz, the dynamic range is in excess of 120 dB and the slew rate – the maximum rate at which the flux-locked loop (FLL) can track changes in flux without losing lock – is 10^6–$10^7\Phi_0$/s. For high-T_C SQUIDs, a typical white flux noise is a few $\mu\Phi_0$ Hz$^{-1/2}$, with an onset of $1/f$ noise below 10 Hz. The flux modulation scheme eliminates low-frequency $1/f$ noise due to noise in the current bias and to the components of noise in the critical current of the junctions that fluctuate "in-phase." However, this scheme does not eliminate "out-of phase" critical current $1/f$ noise in the junctions that induces current noise around the SQUID loop, and hence a flux noise in the SQUID. Fortunately, this noise component can be eliminated by one of several methods in which the bias current is periodically reversed. This latter scheme is rarely implemented for low-T_C SQUIDs where the critical current $1/f$ noise is low, but is essential for high-T_C SQUIDs where it is relatively high. Several other read-out schemes are in use, the most widely used is the direct read-out scheme [33], in which the voltage across the SQUID is coupled directly to the room-temperature amplifier. A positive feedback scheme – in which part of the flux-induced change in voltage across the SQUID is fed back inductively into the SQUID – increases V_Φ for flux between $n\Phi_0$ and $(n + 1/2)\Phi_0$, so that the preamplifier noise is relatively unimportant. This electronics package involves fewer components than is required for the flux-modulated case; bias reversal can be incorporated to reduce the $1/f$ noise in high-T_C SQUIDs. Modern SQUID controllers – especially those for large numbers of SQUIDs – are often equipped with microprocessors that automatically optimize the working point of each SQUID, or reset it if the flux-lock is lost due to momentary overload. Increasingly, the feedback electronics is becoming digital. The signal output and control inputs are usually digitized, enabling the user to operate the system from a computer.

5.2.2.1 Practical Devices

A DC SQUID consists of a superconducting loop with inductance L_{SQ} interrupted by two Josephson junctions. Nowadays SQUIDs are typically fabricated in thin film technology, rather than the bulk material SQUIDs that were used in the beginning. Let us now consider one of the simplest designs: the square washer SQUID. Therein the SQUID inductance is shaped as a washer with inner and outer dimensions d and D.

Although these "bare" or uncoupled SQUIDs, as no external signal other than the flux threading the hole is coupled to the SQUID, have small inductances given by $L_{SQ} \approx \mu_0 d$, they exhibit a very small effective flux capture area of about $A_{eff} = \partial\Phi/\partial B = dD$ [34] and therefore poor magnetic field noise

$$\sqrt{S_B(f)} = \frac{\sqrt{S_\Phi(f)}}{A_{eff}} \tag{5.13}$$

They are favorable for applications where a good spatial resolution is needed, like in SQUID microscopy or miniature susceptometers [35].

To increase the effective area of these devices without changing their inductance one can simply increase the outer dimension D of the washer, and make use of the flux-focusing effect due to perfect diamagnetism in superconductors. Although this method has successfully been applied especially in high-temperature superconductor devices, the increased line width w of the superconductor may deteriorate the low-frequency performance due to trapped flux. A more effective approach is to place a multi-turn thin film input coil on top of the SQUID washer to ensure a tight inductive coupling between both. These two layers are separated from each other by an insulating layer. Now a separate pickup loop with much larger effective area can be connected to this input coil to improve the magnetic field resolution. In addition to the input coil, a second coil is typically integrated on top of the SQUID washer in order to couple a feedback signal to the SQUID. By integrating a thin film input coil, SQUIDs can be implemented not only as SQUID magnetometer, but also as sensors for any physical quantity that can be transformed into magnetic flux. Magnetic

Table 5.1 Typical sensitivities of SQUID sensors for some of applications [26].

Measurement	Sensitivity
Magnetic field	10^{-15} T/Hz$^{1/2}$
Current	10^{-13} A/Hz$^{1/2}$
Voltage	10^{-14} V/Hz$^{1/2}$
Resistance	10^{-12} Ω
Magnetic moment	10^{-10} emu

gradiometers, current sensors and voltmeters, susceptometers, (rf) amplifier or displacement sensors are possible implementations. SQUIDs are therefore very versatile and their applications range from biomagnetism [7, 36] and geophysical exploration [7, 37] to magnetic resonance imaging (MRI) [38]. Typical sensitivities of SQUIDs for some of these application scenarios are listed in Table 5.1.

For the comparison of SQUIDs with different inductances L_{SQ}, one usually refers to the equivalent energy resolution ε. As this describes the energy resolution of an uncoupled SQUID, in practice the so-called coupled energy resolution ε_C is used, which is given by

$$\varepsilon_c = \frac{\varepsilon}{k_{in}^2} \tag{5.14}$$

with k_{in} being the coupling constant between the input coil inductance L_{in} and the SQUID loop inductance L_{SQ}. It is determined via the mutual inductance M_{in} between the input coil and the SQUID

$$M_{in} = k_{in}\sqrt{L_{SQ}L_{in}} \tag{5.15}$$

According to the definition of ε, the coupled energy resolution ε_C corresponds to the minimum energy that can be detected in the input coil per unit bandwidth. Depending on the intended application and accordingly L_{in}, present SQUIDs exhibit coupled energy resolutions of below 100 h.

The design of coupled SQUIDs on the base of washer SQUIDs is rather straightforward and can easily be carried out based on experimentally proven expressions [39, 40]: The SQUID inductance is thus given by

$$L_{SQ} = L_h + L_S + L_j \tag{5.16}$$

Here L_h is the inductance of the washer hole, L_S the inductance of the slit, and L_j the inductance associated with the Josephson junctions (which typically can be neglected). The inductance of the washer hole is

$$L_h = \alpha\mu_0 d \tag{5.17}$$

with $\alpha = 1.25$ for square washer, $\alpha = 1.05$ for an octagonal washer, and $\alpha = 1$ for a circular washer. The slit inductance can be approximated by

$$L_s = \frac{0.3pH}{\mu\alpha} \tag{5.18}$$

The mutual inductance between the SQUID and an integrated multi-turn input coil on top of the SQUID washer is given by

$$M_{in} \approx nL_{SQ} \tag{5.19}$$

The inductance of the input coil can be expressed as

$$L_{in} \approx n^2 L_{SQ} \tag{5.20}$$

The input coil inductance should be matched to the inductance of the pickup circuit for optimum coupling. In case of a SQUID magnetometer the pickup loop is typically a thin film or wire wound loop with inductance L_p. The shape of the pickup circuit has to be adapted to the measurement task. In practical devices, however, deviations from the ideal behavior may appear, as, e.g. stray capacitances between the SQUID washer and the input coil can lead to resonances in the flux-voltage characteristics and may therefore strongly deteriorate the device performance. In consequence, a careful design optimization procedure is typically required for such tightly coupled SQUIDs. Detailed information on this topic can be found, e.g. in [41, 42]. Although the coupled energy resolution is a good method to compare SQUIDs with different inductances, the figure of merit for, e.g., a magnetometer is the magnetic field noise $S_B^{1/2}$. It has been shown that the magnetic field resolution and hence the magnetic field noise $S_B^{1/2}$ can be improved by increasing the pickup loop area while maintaining $L_{in} \approx L_{in}$. In [25] the approximation for the white noise level of $S_B^{1/2}$ vs. the radius r_p of the pickup loop is given as

$$\sqrt{S_B} \approx \frac{2\sqrt{\mu_0\varepsilon}}{r_p^{3/2}} \tag{5.21}$$

For a circular pickup loop area with $r_p = 15$ mm and $\varepsilon = 10^{-32}$ Js this results in $S_B^{1/2} \approx 100 \times 10^{-16}$ T/Hz$^{1/2}$ = 0.1 fT/Hz$^{1/2}$. It is worth to note at this point that although the SQUID sensor by itself may exhibit such an excellent noise performance, the overall noise performance of the SQUID system may be impaired by, e.g. noise of the readout circuit as well as the environment as, for example, by noise arising from the dewar. Another way to increase the effective area of the magnetometer while maintaining the SQUID inductance at a tolerable level is to divide the superconducting pickup loop into a number of separate loops connected in parallel in order to reduce the total SQUID inductance. In these so-called multiloop magnetometers, as described in detail in [43–46], the SQUID itself typically acts as the sensitive area, whereas the so-called Ketchen-type SQUID is inductively coupled to an antenna as discussed above. Due to inevitable losses owing to the used flux transformer in inductively coupled SQUIDs, the multiloop magnetometer allows for the best field resolution for a given chip area, as for instance described for first-order gradiometers in [47].

However, transformer-coupled SQUIDs offer the possibility to include thin-film low-pass filters in the design to increase their robustness – especially for electromagnetically unshielded operation. The SQUID itself is shaped in form of a clover leaf with the input coil on top. The layout of the SQUID as a first-order gradiometer results in its insensitivity to homogenous ambient field and it may thus be operated as a current sensor.

5.2.3 rf SQUID

The rf SQUID, shown in Figure 5.5, employs only a single junction in a super-conducting loop. The loop inductance L is coupled to the inductor L_T of a tank circuit via a mutual inductance $M = K(LL_T)^{1/2}$. The tank circuit is driven by a current oscillating at or near the resonant frequency, $f_0 = \omega_0/2\pi$, which may vary from 20 MHz to 10 GHz. The resistance R_T represents the loss in the tank circuit, so that the unloaded quality factor is $Q_0 = \omega_0 L_T/R_T$ in the absence of the SQUID. On resonance, and with the SQUID in place, the oscillating bias current $I_{rf}\sin\omega_{rf}t$ thus induces a current $I_T\sin\omega_{rf}t = I_{rf}\sin\omega_{rf}t$ in the inductor; here Q is the loaded quality factor. The peak rf flux in the SQUID loop is $\Phi_{rf} = MI_T$. The tank circuit, which is connected to a preamplifier, also serves to read out the applied flux Φ_a in the SQUID: the amplitude of the rf voltage $V_T\sin\omega_{rf}t$ is periodic in Φ_a with period Φ_0.

We briefly describe the behavior of the rf SQUID [48–51]. It is assumed that the junction is damped sufficiently to eliminate hysteresis. Flux quantization imposes the constraint

$$\delta + \frac{2\pi\Phi_T}{\Phi_0} = 2\pi n \tag{5.22}$$

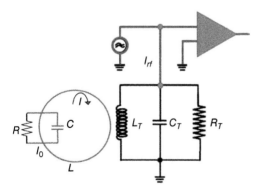

Figure 5.5 Schematic representation of the rf SQUID, with tank circuit and preamplifier; the operation point is set by the amplitude of the rf bias current, I_{rf}. Source: adapted from [25].

on the total flux Φ_T in the loop, where n is an integer. In turn, the phase difference δ across the junction determines the supercurrent

$$J = -I_0 \sin \left(\frac{2\pi\Phi_T}{\Phi_0} \right) \tag{5.23}$$

flowing around the loop. The total flux can thus be written

$$\Phi_T = \Phi_a - LI_0 \sin \left(\frac{2\pi\Phi_T}{\Phi_0} \right) \tag{5.24}$$

Inspection of Eq. (5.24) shows that there are two distinct kinds of behavior.
For $\beta_{rf} \equiv 2\pi LI_0/\Phi_0 < 1$, the slope $d\Phi_T/d\Phi_a = 1/[1 + \beta_{rf}\cos(2\pi\Phi_T/\Phi_0)]$ is everywhere positive and the Φ_T vs. Φ_a curve is nonhysteretic. On the other hand,
for $\beta_{rf} > 1$ there are regions in which $d\Phi_T/d\Phi_a$ are positive, negative, or diver-
gent, so that the device makes transitions between flux states. As a result, the Φ_T
vs. Φ_a plot is hysteretic. The rf SQUID may be operated in either regime. For
high thermal fluctuations ($\Gamma > 1$), the nonhysteretic regime extends to higher
values of β_{rf} predicted and experimentally verified to be about 3 [52, 53]. The
hysteretic regime for $\Gamma > 1$ has not yet been analyzed.

In the hysteretic mode, the rf drive causes the SQUID to make transitions
between quantum states and to dissipate energy at a rate that is periodic in
Φ_a; hence, this is termed the "dissipative mode." This periodic dissipation in
turn modulates the Q of the tank circuit, so that when it is driven on resonance
with a current of constant amplitude, the rf voltage is periodic in Φ_a. The characteristic "steps and risers" are evident, as is the change in amplitude of V_T with
Φ_a at appropriate values of I_{rf}. The nonzero slope of the steps is due to thermal
noise. A detailed analysis shows that the device is optimized when [48–51]

$$k^2 Q \approx 1 \tag{5.25}$$

at which value the flux-to-voltage transfer coefficient becomes

$$\frac{\partial V_T}{\partial \Phi_a} \approx \omega_{rf} \left(\frac{QL_T}{L} \right)^{1/2} \approx \frac{\omega_{rf}(L_T/L)^{1/2}}{k} \tag{5.26}$$

We note that V_Φ scales as ω_{rf} and as $(1/L)^{1/2}$. The *intrinsic* noise energy is
given by [54]

$$\varepsilon(f) \approx \frac{LI_0\Gamma^{4/3}}{2\omega_{rf}}. \tag{5.27}$$

However, for rf SQUIDs at 4.2 K and a room-temperature preamplifier, it
should be stressed that there are extrinsic contributions to the noise energy that
may far exceed the intrinsic value. These contributions arise in part because the
noise temperature of the rf preamplifier tends to be significantly higher than
the bath temperature and in part because portions of the coaxial line connecting

the tank circuit to the preamplifier are at room temperature. The latter contribution can be eliminated by cooling the preamplifier; however, this is an additional complication. Furthermore, one needs to operate the SQUID at high
values of ω_{rf} – perhaps as high as $R/10L$ [54] – to achieve an intrinsic noise
energy comparable to that of the DC SQUID at 4.2 K. It is for these reasons that
rf SQUIDs are rarely used today at liquid helium temperature.

A corollary to this behavior, however, is that when one increases the operating
temperature of an rf SQUID coupled to a room-temperature amplifier from 4.2
to 77 K, the system noise energy is virtually unchanged. For the DC SQUID,
however, the noise energy scales with the temperature for fixed SQUID parameters, and the noise advantage of the DC SQUID over the rf SQUID is
diminished.

In the nonhysteretic regime ("dispersive mode") the rf SQUID behaves as a
flux-sensitive inductor [48, 50–52, 55, 56]. When the tank circuit is operated
off-resonance, a flux change in the SQUID thus changes the resonant frequency,
causing the rf voltage across the tank circuit to be periodic in Φ_a at constant
drive amplitude and frequency. The "steps and risers" in the plot of rf voltage
vs. current are qualitatively similar to those for the dissipative mode. However,
when $kQ^2\beta_{rf} > 1$, that is, when the tank circuit is rather strongly coupled to the
SQUID and Q is high, the resonance curve $V_T(f)$ is nonlinearly dependent on Φ_a,
becoming asymmetrical and even multivalued. Consequently, when the bias frequency is set near a point of infinite slope, the transfer coefficient can be arbi-
trarily high, but at the cost of reduced bandwidth and strong dependence on
SQUID parameters. In practice, when the detuning from resonance is appropri-
ately chosen, a very high value of $\partial V_T / \partial \Phi_a$ can be obtained with wide bandwidth
and acceptable stability. In the low fluctuation limit, $\Gamma \ll 1$, the intrinsic noise
energy can be approximated by [52]

$$\varepsilon(f) \approx \frac{3k_B TL}{\beta_{rf}^2 R} \tag{5.28}$$

The intrinsic noise energy may be lower than in the dissipative mode. Furthermore, in contrast to the dissipative mode, the noise of a room-temperature preamplifier and coaxial line may not dominate, even for a 4.2 K SQUID. This
becomes possible when the SQUID is interacting strongly with the tank circuit,
i.e. $k^2 Q\beta_L \geq 1$. It should be noted the lowest flux noise and energy resolution of
any rf SQUID were reported for a microwave SQUID operating in this regime
[57]. The intrinsic noise energy of dispersive SQUIDs remains low even in the
high fluctuation limit, making it possible for them to operate with relatively
large loop inductances, somewhat higher than those of DC SQUIDs. Large
inductances, and correspondingly large loop areas, help one to attain high field
sensitivities in single-layer high-T_C SQUIDs.

Almost all practical high-T_C rf SQUIDs are made from a single layer of YBCO.
The grain boundary junctions are formed either on a bicrystal or more often on

a step edge. Most of the structures used to improve the magnetic field sensitivity – for example, by flux focusing – are also fabricated from single-layer YBCO films. Early devices were operated with lumped tank circuits consisting of a wire-wound coil and a capacitor; however, this technology is limited to perhaps 200 MHz. Recent, more sensitive designs involve microwave resonators at frequencies of 0.5–1 GHz. For practical reasons, higher frequencies are rarely used. A multiturn flux transformer has been demonstrated [58] but not implemented in practical applications.

Although cooled preamplifiers have been explored, current practice uses room temperature amplifiers, since the advantages of cooling them to 77 K are relatively modest. After amplification, the rf signal is usually rectified and smoothed to give a quasistatic voltage that is periodic in the flux applied to the SQUID. The rf SQUID is almost invariably operated in a FLL, with flux modulation as for the DC SQUID. It is noteworthy that the combination of rf bias and flux modulation eliminates $1/f$ noise due to fluctuations in the critical current of the junction, analogously to the combination of flux modulation and bias reversal for the DC SQUID. Typical values of white flux noise in high-T_C rf SQUIDs are somewhat below $10^{-5}\Phi_0$ Hz$^{-1/2}$; furthermore, the $1/f$ flux noise is higher than that of single-layer DC SQUIDs operating with bias reversal.

5.2.4 Cryogenics and Systems

Low-T_C and high-T_C SQUIDs are traditionally operated at liquid He temperatures (≤ 4.2 K) and liquid nitrogen temperatures (≤ 77 K), respectively [52]. For some applications the SQUID and its associated circuitry are immersed in the cryogen, which is contained in a suitable dewar made of glass, metal, or (most commonly) fiberglass. In other applications, it is more convenient to operate the device in a vacuum, with conductive cooling. In recent years, some commercial systems make use of a cryo-cooler, thus obviating the need to replenish the liquid cryogen: this is an important step in improving the user-friendliness of SQUIDs. SQUIDs are often encapsulated to protect them against environmental damage, and provided with a heater to expel trapped flux. Broadly speaking, there are two kinds of SQUID applications: those in which the signal is generated at low temperature and those in which the signal source is at room temperature. In the former case, the entire experiment – SQUID, input circuit and sample – is usually surrounded by a superconducting shield to exclude ambient magnetic field fluctuations. If a low static field is desirable, the Earth's field may be attenuated by surrounding the dewar with one or more shields made from a high permeability material such as mu-metal. In the latter case, the SQUID or the pickup loop of an appropriate flux transformer must obviously be exposed to the signal source outside of the dewar. When a flux transformer is used, the SQUID itself may be enclosed in a superconducting shield with leads connecting its input coil to the flux transformer. For such measurements, it is vital that the dewar and its components generate very low levels of magnetic noise. Furthermore, to maximize the signal, the room-temperature source must be brought as close as possible to the cooled dewar. This requirement calls for a dewar with thin walls and minimal separation between the inner (cold) and outer (warm) wall. The necessary reduction in the number of layers of superinsulation results in a higher boil-off rate of the cryogen or an increased load on the cryo-cooler. In some cases, notably high-T_C "SQUID microscopes," the SQUID and sample may be separated by a single window only a few micrometers thick. Given the extraordinarily high sensitivity of SQUIDs, it goes without saying that suppressing environment noise is a major undertaking. There are two broad issues. The first is magnetic noise in the signal bandwidth, most notably from power lines at frequencies of 50 or 60 Hz and their harmonics. These disturbances can be suppressed substantially by the use of spatial gradiometers, but a high-permeability enclosure for the cryostat and, for example, a human subject may be required when the signals are weak. Such magnetically shielded rooms (MSRs) usually incorporate both high-permeability shields and eddy-current shields made of high-conductivity aluminum. The second – and often even more insidious issue – is radio frequency noise from radio and televisions stations, and particularly from nearby computers and other digital equipment which produce copious levels of electromagnetic interference. Elimination of such noise is a skilled art: the essential principle is to surround the cryogenic components, the leads connecting them to the room-temperature electronics and the electronics themselves with a Faraday cage. In the case of DC SQUIDs, the leads are preferably twisted pairs inside metal tubes, whereas the higher frequencies associated with rf SQUIDs demand coaxial cable – which has a higher thermal loss than a twisted pair.

5.2.5 SQUID Electronics

The SQUID itself acts as a very sensitive magnetic flux-to-voltage transducer with nonlinear periodic flux-to-voltage characteristic. In order to obtain a linear dependence of the voltage across the SQUID from the flux threading the SQUID loop, the SQUID is operated in a feedback loop called FLL.

5.2.5.1 Flux Locked Loop

There are two main FLL schemes [59]: flux-modulation and directly coupled readout. Due to its ability for the design of compact readout circuits, which are suitable for the use in multichannel systems with a sufficiently large bandwidth and dynamic range as well as lower power consumption, the directly coupled SQUID electronics is typically used nowadays [26].

Before going into details of the directly coupled readout, it should be mentioned that with the flux-modulation readout scheme the preamplifier

Figure 5.6 Schematics of a directly coupled SQUID electronics. R_{Fb} and M_{Fb} denote the feedback resistor and mutual inductance between feedback coil and SQUID, respectively. In feedback mode the output voltage V_{out} is linearly dependent on the external signal flux Φ_{sig}. *Source:* adapted from [26].

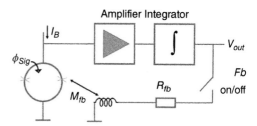

low-frequency noise and in-phase critical current fluctuations of the Josephson junctions are suppressed. As critical current fluctuations in state-of-the-art LTS tunnel junctions are generally very weak, this is not a major concern for most applications. There are as well readout options like bias reversal [60, 61], which allow suppressing in-phase and out-of-phase critical current fluctuations in both readout schemes. The directly coupled readout scheme is schematically shown in Figure 5.6. The voltage across the SQUID due to a changing signal flux Φ_{sig} is amplified, integrated and fed back to the SQUID as a feedback flux Φ_{Fb} via a feedback resistor R_{Fb} and a mutual inductance M_{Fb}.

The FLL therefore keeps the flux inside the SQUID constant and the output voltage, the voltage across the feedback resistor, becomes linearly dependent on the applied signal Φ_{sig} with a strongly increased linear working range. Besides the linearization, the main purpose of the electronics is to read out the voltage across the SQUID without compromising the low voltage noise level of the SQUID. The influence of the read-out electronics on the total measured flux noise $S_{\Phi,t}^{1/2}$ can be expressed as [59]

$$\sqrt{S_{\Phi,t}} \approx \sqrt{\left(\sqrt{S_{\Phi,SQ}}\right)^2 + \left(\frac{\sqrt{S_{V,Amp}}}{V_\Phi}\right)^2 + \left(\frac{\sqrt{S_{I,Amp}}+R_{dyn}}{V_\Phi}\right)^2} \qquad (5.29)$$

Here $S_{\Phi,SQ}^{1/2}$ is the intrinsic flux noise of the SQUID, $S_{V,Amp}^{1/2}$ and $S_{I,Amp}^{1/2}$ are the preamplifier input voltage and current noise, respectively. R_{dyn} denotes the dynamic SQUID resistance in the working point. Typical input voltage and input current noise of state-of-the-art SQUID electronics are about 0.35 nV/Hz$^{1/2}$ and 2–6 pA/Hz$^{1/2}$ [62, 63]. For currently available DC SQUIDs the usable voltage swing and transfer function can typically vary between 30–150 μV and 100–500 μV/Φ_0, respectively. The dynamic resistance of such SQUIDs is usually between 5 and 50 Ω. As a result, the contribution of the room-temperature SQUID electronics can amount up to 1–5 $\mu\Phi_0$/Hz$^{1/2}$ and may thus considerably contribute to the total measured flux noise. Note that the expression above does not account for the noise contribution due to thermal noise in the feedback resistor, given by $S_I^{1/2} = (4k_BT/R_{Fb})^{1/2}$. This current noise converts into flux noise in the SQUID via the mutual inductance M_{Fb}. Especially in SQUID systems requiring a large dynamic range, for example, for unshielded operation within the Earth's magnetic field, this noise, however, may become important or even dominant.

Since SQUIDs are vector magnetometers, a rotation in the Earth's field results in a field difference of up to 130 µT. Thus a SQUID magnetometer system with magnetic field noise of, for example, 10 fT/Hz$^{1/2}$ would require a dynamic range of the order of 200 dB which is larger than 30 bit. Even if the SQUID electronics would allow such an operational range, current analogue to digital converters (ADC) are nowadays still limited to about 24 bit. Besides the dynamic range, another important parameter correlated with the dynamic behavior of the FLL is the system slew rate [59, 62] given by

$$\Phi_{max} = \left|\frac{\partial\Phi_{Fb}}{\partial t}\right| = 2\pi \cdot f_{GBP} \cdot \delta V \cdot \frac{M_{Fb}}{R_{Fb}} \tag{5.30}$$

It describes the maximum signal change in a certain time interval that the electronics is able to follow. Here f_{GBP} is the gain-bandwidth product, a fixed value for a specific amplifier configuration and δV describes the usable voltage swing of the SQUID. Accordingly, a high system slew-rate demands a large δV and a small feedback resistor value, which however may limit the system noise. The configuration of the feedback circuit is therefore always a trade-off between low system noise and high dynamic range and slew rate.

5.3 SQUID Fabrication

The fabrication of LTS SQUIDs is based on sophisticated thin-film techniques similar to their use in semiconductor industry. SQUID sensors are fabricated on wafers, which are then diced into chips with dimensions of several mm^2 size depending, e.g. on the necessary pickup area for the envisaged application. Quartz, silicon, or oxidized silicon wafers sized 4 inch or larger are typically used as substrates. Therefore, hundreds of SQUIDs can be fabricated in one run. In this section, the basic thin-film techniques used for the fabrication of LTS SQUIDs will be commented and the most important step, junction fabrication, will be highlighted. More detailed information can, e.g. be found in [64].

5.4 Lithography and Thin-Film Techniques

Nowadays superconducting thin film materials for LTS SQUIDs are mainly Nb and Al. In the beginning, usually Pb or Pb alloys have been used (as well as electrode material for the junction fabrication), but the limited long-term stability and problems associated with thermal cycling have led to the "all-refractory" process used today. To fabricate thin superconducting films, various deposition techniques such as thermal or e-beam evaporation, molecular beam epitaxy, plasma, and ion beam sputtering can be used. Due to the high melting temperature of Nb, sputtering is de facto the standard. This is typically done in ultra-high vacuum, as impurities may dramatically change the superconducting thin-film properties.

A careful optimization of the deposition and patterning process of superconducting films with respect to their influence on, e.g. minimum film stress, superconducting properties or the shape of the structured edges is essential. Steep edges of superconducting films are usually favorable, as they are less susceptible to flux trapping. In multilayer processes, moreover, special attention has to be paid to avoid residues or fence structures associated with the patterning of the films as they may lead to shortcuts in or failure of the devices. Higher integrated multilayer processes like the Josephson junction-based rapid single flux quantum (RSFQ) logic [65, 66] try to overcome difficulties associated with an increased number of superconducting layers and therefore potential step height or surface topography problems by planarization of isolation layers (typically with chemical mechanical polishing). As the design of SQUIDs is usually less complex than RSFQ circuits, planarization is in generally not performed in SQUID fabrication nowadays, but it may be implemented in future.

The patterning of the thin films is either done by lift-off or by etching. For lift-off, the photoresist is applied to the substrate prior to the thin film deposition. When the thin film is patterned via etching, the photoresist is placed on top of the thin film. In both cases the resist acts as a mask for the structure to be defined. For lift-off the resist is removed in an (ultrasonic) solvent bath so that the film on top of the resist is removed as well. The etch process is typically done by dry etching such as plasma or reactive ion-beam etching. Wet etching may as well be used, but is not that attractive due to the isotropic etch behavior. To avoid over-etching of the underlying film, one can either use an end-point detector or make use of thin natural etch stops like an Al layer for a fluorine-based etch process.

In general, elevated temperatures should be avoided (especially when the trilayer to form the Josephson junctions is already deposited on the wafer), as this increases, e.g. the diffusion of hydrogen into the thin film or may change the barrier characteristics [67, 68].

The typical film thickness is in the range of 50 to about 300 nm. The line width of superconducting structures such as patterned multi-turn input coils on top of the SQUID washer may be as small as 1 μm or even less. The resist thickness depends on the lithography method and the lateral dimension of the desired thin film structure and may vary between several hundred nm to about 2 μm. Depending on the design complexity, the fabrication of LTS SQUIDs includes at least two superconducting layers, one for the SQUID washer and one for the input and feedback coils, and (several) isolation layers.

5.4.1 Junction Fabrication

Nowadays SIS Josephson junctions are typically based on a sandwich of an in situ deposited $Nb - AlO_X - Nb$ trilayer. There are other material systems like, e.g. $Nb - SiN_X - Nb$ but they do not exhibit such a good junction quality, reproducibility, low junction capacitance, and low level of critical current fluctuations. Detailed information on other material systems used in the past can be found in [7, 64]. Today, most fabrication technologies are based on the so-called SNAP process (selective niobium anodization process) [69] or its numerous variations. In 1983, Gurvitch introduced the use of $Nb - AlO_X - Nb$ Josephson junctions [70]. This material combination has led to superior junction characteristics and became soon the most important junction fabrication process. Up to now it is the standard even for very complex RSFQ circuits for digital applications and it allows the reliable fabrication of up to tens of thousands Josephson junctions on a single chip [65].

The junction fabrication starts with the deposition of a trilayer consisting of an Nb base electrode, a thin Al layer (which is partly oxidized during the trilayer deposition) and another Nb layer as counter electrode. The in situ deposition of the trilayer is essential for clean interfaces between these layers. The AlO_X is formed by exposing the sputtered Al to pure oxygen atmosphere for a certain time. The thickness of the AlO_X layer t_{ox} – given by the product of oxygen partial pressure and exposure time – determines the junction's critical current density j_C, which is exponentially dependent on t_{ox}. For SQUIDs j_C is in the range of $0.1–2$ kA/cm^2 depending on the desired junction's critical current and size. The typical film thickness is 50–300 nm for Nb layers and about 10 nm for Al. The thin Al layer is used to level out the surface roughness of the underlying Nb layer and allows a low junction capacitance due to the much lower dielectric constant ε_r of AlO_X compared to NbO_X.

In the SNAP process the junction area is defined by anodizing the upper electrode of the trilayer. During anodization the desired junction area is covered by a small resist dot. In this so-called window-type process the typical minimum junction size is several μm^2. Since the anodization solution creeps partly under the photoresist, small junctions are less reproducible or even defective.

For electrical connection of the junction, a Nb layer is deposited on top of the counter electrode. Finally, a shunt resistor is placed close to the junction to damp its dynamics and to fulfill the condition $\beta_C \leq 1$. Usually Pd, AuPd, Ti, or Mo is used as shunt material. The specific capacitance of a $Nb - AlO_X - Nb$ Josephson junction (it forms a parallel-plate capacitor) is about 45–60 fF/ μm^2, depending on the barrier thickness and therefore on the critical current density [71]. Due to the overlap of superconducting layers around the junction (e.g. to compensate inevitable alignment errors between different layers), a parasitic capacitance is formed, which adds to the junction capacitance. The influence of this effect becomes even more pronounced as the junction size is reduced.

A small total junction capacitance is favorable since it will improve the performance in terms of energy resolution and voltage swing of the SQUID. To reduce or even avoid parasitic capacitance, several fabrication technologies have been reported. One possible approach is the so-called cross-type technology [72], in which the junction is defined by the overlap of two narrow perpendicular strips. The lower strip is the entire Nb – AlO$_X$ – Nb trilayer, which is patterned with the width corresponding to the desired linear dimension of the junction. The second perpendicular strip of Nb is deposited on top of the trilayer and acts as a mask for patterning the Nb counter electrode from the trilayer. Due to the self-alignment of the process, no parasitic capacitance is formed. In [72], high-quality Josephson tunnel junctions with dimensions of $0.6 \times 0.6 \ \mu m^2$ have been reported. Due to the narrow linewidth design of the junctions, flux trapping is avoided and these devices can be cooled in the Earth's magnetic field without restrictions [73]. The current trend in superconducting fabrication technology is the further decrease in junction capacitance and accordingly a downsizing of the Josephson junctions, while maintaining a high fabrication yield and low parameter spread over the entire wafer.

5.5 SQUID Applications in Biomagnetism

5.5.1 Biomagnetism

Electrical activity in the living body is caused by movements of ions inside, outside and across cellular membranes [7]. These movements of electrically charged particles, natural electrical currents, are responsible for magnetic fields measurable outside the body, which are called *biomagnetic fields* [74]. Examples of such ion currents in humans are those from

1) myocardial activity, which produce the magnetocardiogram (MCG)
2) the neuronal currents in the head produce parts of the magnetoencephalogram (MEG)
3) the currents in the unborn child (heart or brain) generate the fetal magnetocardiogram (fMCG) or fetal magnetoencephalogram (fMEG).

The measurement of the magnetic fields produced by the human body is called the biomagnetic measurement technique and is a completely noninvasive and contact-free method without any influence on the subject. This method is useful for obtaining both spatial (in the millimeter range) and temporal (in the millisecond range) information about magnetic field distribution. Based on the magnetic field measurements and the field distribution, it is possible to localize the sources of the magnetic field, which is used in magnetic source imaging (MSI). All tissues of the human body are practically nonmagnetic. Therefore, the propagation of the magnetic field is not disturbed by human tissue. This represents one principal advantage of magnetic measurement over conventional measurement of potential differences by surface electrodes (electrocardiogram [ECG], electroencephalogram [EEG]), because these potential differences are strongly influenced by conductivity inhomogeneities inside the body.

However, biomagnetic fields are very weak. Therefore, very sensitive magnetic field detectors are necessary and, additionally, disturbances have to be reduced sufficiently in order to achieve a suitable signal-to-noise ratio (SNR). Disturbances may come from the environment (external noise), from the measurement setup and from the subject to be investigated itself, respectively.

In order to localize the sources of the biomagnetic field, it is necessary to make assumptions about the structure of those sources. The simplest and most common assumption is to describe the source as a current dipole. Primary currents denote active, impressed (charge transport caused by chemical processes or concentration gradients) and passive ohmic currents within or in the close surrounding of electrically active cell populations, which are lumped together to a single equivalent current for modeling purpose. This current is often represented by an electric current dipole. The extracellular currents (also called volume currents or secondary currents) are produced by the effects of volume conduction in the tissue surrounding the dipole.

5.5.2 History of SQUID Applications in Biomagnetism

The Philips Company had designed and manufactured a twin-dewar biomagnetometer system [75] with 2×31 channels for operation inside a shielded room. The imbalance of the gradiometers was less than 0.1%, and the field noise of the system was less than $10 \, fT/Hz^{1/2}$ at 1 Hz [76]. The investigations performed with this device could also be used for MSI. With respect to the physiology of the source, different source models can be applied in order to perform MSI [77]. The principle of localizing, for example, an anterior infarction in cardiomagnetism, is based on the same procedure [78]. All these systems belong to the history (for more information, see [76]). The most sensitive noncommercial multichannel biomagnetic measurement system in use has been reported by the Physikalisch-Technische Bundesanstalt (PTB)-group [79–81] with a typical system white noise level of less than $2.3 \, fT/Hz^{1/2}$. Nowadays, the multichannel SQUID-biomagnetometers with hundreds of channels are made for MEG investigations (helmet-shaped systems), for MCG investigations (plane devices) or for fetal magnetography (adapted to the abdomen of pregnant women) due to the different applications.

5.5.3 Biomagnetic Fields

Biomagnetic signals are extremely weak in comparison with the earth's magnetic field or disturbances caused by urban noise. These weak biomagnetic fields are in the order of picotesla and femtotesla, at frequencies from a fraction of one hertz to kilohertz. The strongest field is generated by the human heart (MCG) and by skeletal muscles (magnetomyogram [MMG]). The amplitude of the QRS-peak in the MCG is typically several tens of picotesla. Neuromagnetic signals (MEG) are much weaker. The largest field intensity of a normal awake brain is due to spontaneous activity. The so-called alpha rhythm, observed over the posterior parts of the head, is about 1 pT in amplitude. Typical evoked fields – somatosensory, auditory, or visually evoked responses – are weaker by one order of magnitude or more, their strengths are only several tens or hundreds of femtotesla. Biomagnetic fields are also known from other electrically active organs: the eye as the magneto-oculogram (MOG) and the magnetoretinogram (MRG), the stomach as the magnetogastrogram (MGG), the fetal heart and brain (FMCG or FMEG), and the peripheral nerve as the magnetoneurogram (MNG).

In Figure 5.7, an overview is given of the biomagnetic fields and of ambient magnetic field disturbances as well as the magnetometer resolution. The

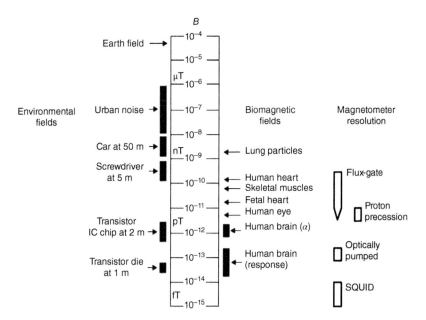

Figure 5.7 Biomagnetic fields and environmental magnetic field disturbances as well as the magnetometer resolution. *Source:* adapted from [82].

magnetic noise from the environment is four to six orders of magnitude stronger than the biomagnetic fields to be measured. Disturbances from environmental magnetic fields are caused by the earth magnetic field as well as by urban noise. The magnitude of the steady earth magnetic field is about 5×10^{-5} T and low frequency variations of this field are in the order of $10^{-7} - 10^{-8}$ T. The urban noise is in the same order caused mainly by power lines, traffic (e.g. car movements or passing trains) and by vibration. When performing biomagnetic measurements we are faced with a twofold problem: very weak biomagnetic signals have to be measured in the presence of environmental magnetic noise, which is many orders of magnitude stronger than the fields to be detected. Therefore, a very sensitive sensor is needed as well as the capability of reducing the ambient noise below the signal to be measured.

5.5.3.1 Gradiometers

Being superconducting, a flux transformer provides noiseless magnetic gain between the field detected by the pick-up coil, and that seen by the SQUID sensor. In the case of thin film SQUIDs planar input coils are tightly and inductively coupled to the SQUID loop [39, 44]. Coupling factors (k^2) of about 0.8 can be achieved with planar thin film structures. Due to an increasing of the sensor area by a flux transformer the sensitivity of the SQUID system can be improved to several fT/Hz$^{1/2}$.

Most biomagnetic measurement systems use gradiometers to reduce the ambient magnetic noise instead of magnetometers [83] which measure the magnetic field itself at only one point or region. Gradiometers measure the difference in magnetic field between the coils. Biomagnetic fields are very inhomogeneous and decrease as $1/r^2$ to $1/r^3$ with the distance r. Gradiometers are sensitive to inhomogeneous sources placed near the pickup coil and reduce the output for uniform background fields. The principle of a gradiometer is the following: the magnetic flux to be detected penetrates two coils, which are connected in series but wound in opposite directions as shown in Figure 5.8 for an axial superconducting gradiometer.

The induced current caused by a homogeneous magnetic field B (ambient disturbances due to the earth's magnetic field or power lines) will cancel out, whereas an inhomogeneous field B (biomagnetic field) will yield different currents in each coil. Such *first-order gradiometers* do not measure the field itself but the field difference between two points, that is, the field gradient. This is useful because the distance between antenna and the source of the disturbance is large compared with that between pickup coil and biomagnetic field source, so the disturbing fields can be assumed to be nearly homogeneous.

A second-order gradiometer can be built by connecting two first-order gradiometers in series. This type of antenna can suppress homogeneous fields and field gradients. Second-order gradiometers are commonly used for measuring in unshielded environments. A wire-wound gradiometer has in the best case

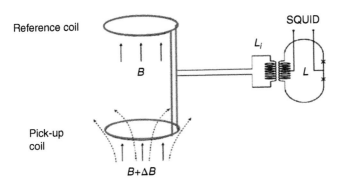

Figure 5.8 Scheme of a first-order axial gradiometer (pickup coil and reference coil) coupled with a SQUID; L_i, inductance of the input coil; L, inductance of the SQUID. *Source:* adapted from [7].

a manufacturing accuracy of nearly 10^{-3}. Such gradiometers may be balanced mechanically [84, 85] to increase their accuracy by about 1000 [86]. Provided that the magnetic field to be measured only depends on the z-direction (axial gradiometer), the conditions for the gradiometer can be given by

$$\sum_{i=1}^{n+1}(-1)^{i+1}n_iA_i = 0 \tag{5.31}$$

and

$$\sum_{i=2}^{n+1}(-1)^{i+1}(z_i-z_1)^k n_iA_i = 0 \tag{5.32}$$

to discriminate the kth field derivative. A_i and n_i are the area and number of turns, respectively, of the single gradiometer coil. $(-1)^{i+1}$ describes the direction and the coefficient $(z_i - z_1)$ the distance between the compensation coil (i) and the pickup coil. Therefore, a gradiometer with the order of n is able to compensate field derivatives to the $(n-1)$th order. Software gradiometers and electronic balancing, respectively, were first reported by Williamson et al. [87]. Software gradiometers can be built up from measurement sensors and reference sensors. They utilize reference magnetometers [18, 88, 89] or reference gradiometers in some cases [90, 91]. Direct feedback, offline subtraction, or adaptive signal processing [92] is applied. Implementing the gradiometers in software seems to be the most practical method for multichannel systems because one reference system can serve a large number of sensors.

The baseline of a gradiometer is the distance between the two coils (pickup coil and reference coil). The baseline has to be chosen in such a way that it is long enough to reduce the effect of the biomagnetic field at the second coil to a relatively negligible level in comparison with the effect at the pickup coil.

In multichannel systems cross talk occurs between the single channels [36, 93], because the biomagnetic field itself causes a current in the flux transformer circuit of every channel. Thus, a magnetic field arises in the vicinity of the flux transformer circuit and causes a magnetic flux in the neighboring channels. The cross talk depends on the mutual inductance between the neighboring channels as well as on the self-inductance of the flux transformer circuit. The cross talk coefficient is defined as the ratio between this induced magnetic flux and the original magnetic flux in the measurement channel. To overcome this type of mutual influence ter Brake et al. [93] recommended external feedback which makes the flux transformer circuit currentless (current-locked-mode) and which is used in most of the multichannel devices.

5.5.4 Magnetoencephalography

Presently, magnetoencephalography (MEG) is the most important biomagnetic application and its technology has been intensely developed in the commercial sector, resulting in complex systems with large numbers of channels covering the whole cortex surface. The first whole-cortex MEG systems [94, 95] were introduced in 1992 by VSM Med-Tech Ltd (then CTF Systems Inc.) and Elekta Neuromag Oy, and later on by 4D Neuroimaging. In addition to these three major suppliers, whole-cortex MEG systems are also manufactured in Japan [96] and in Italy [97]. Noncommercial SQUID magnetometers with large numbers of channels were also constructed in a number of laboratories around the world, using both LTS and HTS. Some examples, not intended to be exhaustive, can be found in [18, 19, 98–106].

The introduction of the helmet-type MEG systems has led to a dramatic increase in the use of SQUID sensors. It was estimated that the number of SQUID sensors installed in commercial whole-cortex MEG systems from 1992 to the end of 2004 was in excess of 20 000, while the total number of functional SQUIDs in all applications before 1992 (a period of 20 years since the first SQUID recording of the human brain magnetic fields in 1972) was only slightly more than 1000 [107].

The existing commercial MEG instruments are based on low-T_c SQUIDs. Systems based on high-T_c superconductors are also being developed, however, mostly for MCG applications [106, 108]. The present high-T_c SQUIDs cannot be reliably manufactured in large quantities, are not as sensitive as their low-T_c counterparts, and require better shielding. However, their performance is steadily improving and they already are suitable for some applications.

The MEG measures magnetic fields on the scalp surface. However, the brain current distribution, which is responsible for the observed fields, is usually more interesting to the user. Unfortunately, the inversion problem (computation of the current distribution from the measured magnetic field) is nonunique and ill-posed and the MEG data must be supplemented by additional information, physiological constraints, or mathematical models and simplifications. Additional information to assist the field inversion can also be supplied by other measuring modalities.

One such modality is electroencephalography (EEG) [109]. Both MEG and EEG measure the same sources of neuronal activity and their information is complementary and additive [110]. Both MEG and EEG have excellent temporal resolution and provide functional information which is usually combined with anatomical images, obtained, for example, from MRI [111, 112] or computed axial tomography (CAT or CT) [111, 113]. Additional functional information from positron emission tomography (PET) [114, 115], single-photon emission computed tomography (SPECT) [115, 116], or functional MRI (fMRI) [115, 117, 118], can also be combined with MEG and EEG to characterize the brain sources more fully.

In summary, MEG and EEG provide a direct measure of the neuronal activity with excellent temporal resolution, but with spatial localization dependent on the nonunique inversion problem. Both MRI and computed axial tomography (CT) provide high-resolution spatial anatomic images and functional magnetic resonance imaging (fMRI), PET, and SPECT provide three-dimensional functional characterization of the brain activity in terms of metabolic and hemodynamic processes. In comparison with MEG and EEG, fMRI, PET, and SPECT are limited by the long time constants of the metabolic and hemodynamic processes and by the poorly defined relationship between them and neuronal processes.

5.5.4.1 MEG Signals

In this section, the origins of MEG signals will be briefly reviewed. More detailed exposition of the cellular mechanism of the magnetic field generation can be found elsewhere [119–122]. The MEG fields are generated by currents flowing within the brain, especially the cortex. The cortex contains well-aligned pyramidal cells, which consist of dendrites, a cell body, and an axon, and there are approximately 10^5–10^6 cells in an area of about 10 mm^2 of the cortex [123]. The cells are interconnected by nerve fibers, which are connected to dendrites and cell bodies by synapses. In the human brain there are approximately 10^{10} neurons and 10^{14} synaptic connections.

The dendrite (or a cell body) can be thought of as a tubular volume surrounded by a membrane. Because of the Na–K pump mechanism [119], there is an excess concentration of K^+ ions inside and of Na^+ ions outside the cell. These concentration gradients and the difference of the membrane permeability for K and Na ions cause diffusion of the positive ions across the membrane. A competition between the electrical and the diffusion forces (Nernst equation [119]) establishes a negative equilibrium potential of about −70 mV within the cell. Cell stimulation (chemical, electrical, or mechanical) can cause alteration of the cell transmembrane potential and can lead to cell depolarization (or hyperpolarization). Because the cell is conductive, the polarization change induces current flow within the cell (intracellular current) and a return current outside the cell through the brain. The transmembrane currents are called the impressed currents, J^i and they drive passive volume currents in the conducting tissues outside the membrane. The contribution of the volume currents to the magnetic field can be expressed as a sum of terms over all surfaces of conductivity discontinuities, including the cell membranes and the macroscopic volume of the brain itself. Summation over the cell boundary terms can be shown to be equivalent to a sum of dipole sources and can be expressed as a cellular-average current dipole density, J^c. From a macroscopic point of view, the combination of J^i and J^c behaves as an effective current source, and it is convenient to call it the primary source of biomagnetic fields, $J^p = J^i + J^c$. The secondary sources then correspond to the terms associated with macroscopic discontinuities, for example, the brain boundary [124]. The goal of magnetic imaging studies is to find J^p from measurements. Since the membrane volume is small, the contribution of J^i to the magnetic field is negligible and the primary sources can be expressed as $J^p \approx J^c$.

The action potentials or axonal currents are usually not observable magnetically, because they consist of propagating depolarization and repolarization regions, equivalent to two, close-spaced, opposing polarity, in-line current dipoles (quadrupolar source) [124, 125], which give rise to magnetic fields with fast spatial decay. Therefore, observed MEG signals are often thought to arise from postsynaptic currents [122].

The postsynaptic dendritic depolarization currents flow roughly perpendicular to the cortex. However, the cortex is highly convoluted with numerous sulci and gyri and, depending on where the cell stimulation occurred, the current flow can be either tangential or radial to the scalp. If the brain were modeled as a conducting sphere, due to symmetry only the tangential currents would produce fields outside the brain, while there would be no contribution from radial currents [126, 127]. This would seem to indicate that the MEG is not sensitive to gyral sources. However, it has been shown on a purely anatomical basis that only 5% of the whole cortical area is within 15° of the radial direction, and that it is the source depth, rather than orientation, that limits the sensitivity of MEG to the electrical activity of the cortical surface [128]. In addition, radial magnetic fields, which are often measured by MEG, are mostly caused by the tangential primary currents, while the electric potentials (or EEG) are affected also by the volume currents.

Current flow within a single cell is too small to produce observable fields outside the scalp. The detectable fields are a result of nearly simultaneous activation of a large number of cells, typically 10^4–10^5 [120]. Generally, the MEG sources are distributed; however, activation of even a large number of cells is often spatially small and can be modeled by a point equivalent current dipole (ECD) [74]. As an example, consider auditory evoked fields (AEF) which are typically caused by sources with an equivalent dipole strength of the order of 20–80 nA m [129].

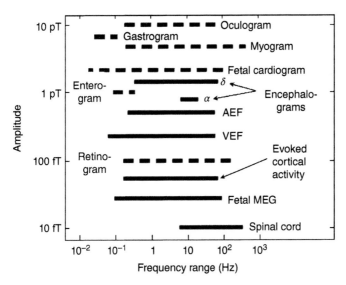

Figure 5.9 Representative biomagnetic fields and their frequency and amplitude ranges. Solid lines indicate brain fields, dashed lines fields from other body parts. *Source:* adapted from [131].

Since the current dipole density in the brain tissue is nearly constant and in the range from about 0.5 to 2 nA m/mm^2 [130], the AEF dipole magnitude corresponds to the order of 100 mm^2 area of activated cortical tissue. Typical ranges of frequencies and amplitudes encountered in MEG are shown by solid lines in Figure 5.9. The field magnitudes range from low 10 fT for the spinal cord to about 1 pT for α or δ rhythms, and in frequency from a fraction of a hertz to about 1 kHz.

5.5.4.2 Sensor Types for MEG

MEG sensing arrays consist of primary sensors and, optionally, references for environmental noise cancellation. The primary sensor flux transformers are located in close proximity to the scalp surface, where they couple to the magnetic field of the brain. The references are located at some distance away from the brain and are designed mostly to detect the environmental noise while being nearly insensitive to the brain signals. The primary sensors are subjected to noise which can be either correlated or uncorrelated among the channels. The correlated noise is either the environmental noise, or noise from electrically active tissues other than brain (for example, muscles [MCG]), or unwanted brain signals. The uncorrelated noise is the random noise generated in the SQUIDs and their electronics.

It will be assumed that all sensor types (with or without the reference noise cancellation) can be designed for the same white noise levels. This section will thus focus on the comparison of the sensor types from the point of view of their response to the brain signals, and their ability to reject the environmental and the brain-generated noise.

Several variations of the primary flux transformers commonly used in MEG practice are shown in Figure 5.10a. Flux transformers (1), (2), and (6) are magnetometers, which have the highest sensitivity to both the near- and far-field sources. This property makes them highly sensitive to the environmental noise and they depend on other techniques for noise cancellation. The flux transformers (3), (4), and (5) in Figure 5.10a are hardware first-order gradiometers. If the gradiometers (4) and (5) were positioned at the same location and if their dimensions were infinitesimally small, they would produce identical signals because outside the head curl B = 0. The gradiometers provide reduced sensitivity to the environmental noise while maintaining good sensitivity to the near-field brain signals. Hardware gradiometers may also require supplemental noise cancellation.

If we disregard both SQUID and environmental noise, the peak signals detected by various primary flux transformers in response to a single ECD

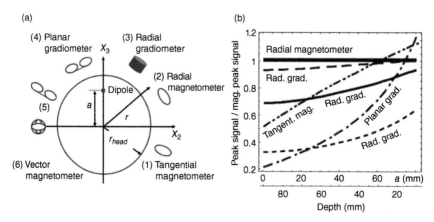

Figure 5.10 Sensor types and comparison of a single ECD source peak signal strengths detected by different noiseless flux transformer types. An ECD source oriented along outward and perpendicular to x_3 and x_2 axes is positioned at a distance a from the model sphere center, sensors are located on a spherical shell with radius $r = 0.11$ m, the head radius is $r_{head} = 90$ mm, and all sensors are assumed to have infinitesimally small coil diameters. (a) Illustration of various sensor types: 1, tangential magnetometer; 2, radial magnetometer; 3, radial gradiometer; 4, planar gradiometer; 5, radial–tangential gradiometer; 6, vector magnetometer. (b) Peak signal normalized by magnetometer peak signal, plotted as a function of the ECD source distance from the model sphere center, or a depth below the head surface: —, radial magnetometers; – - - –, tangential magnetometers; – – –, radial gradiometers $b_{rad} = 0.15$ m; ——, radial gradiometers $b_{rad} = 50$ mm; - - - -, radial gradiometers $b_{rad} = 15$ mm; – – –, planar gradiometers $b_{pln} = 15$ mm. *Source:* adapted from [36].

are shown in Figure 5.10b for a spherical conducting medium [126, 127] as a function of the ECD distance from the model sphere center (or as a function of the ECD depth below the scalp surface) [132]. Since the brain field magnitude is a strong function of the ECD depth, the peak signals have been normalized by the radial magnetometer peak signal to make differences between various flux transformers clearer. As a result, the normalized magnetometer response is unity for all ECD depths.

The radial gradiometer peak signal strength declines with decreasing baseline and increasing ECD depth. For example, in the limit where the dipole source approaches the sphere center (and when the signal for all devices approaches zero), a 50-mm baseline radial gradiometer will detect a signal about 30% smaller than a magnetometer. For more superficial sources the difference is smaller; for example, when the source is about 10 mm deep ($a = 80$ mm), the 50-mm baseline radial gradiometer signal is only about 5% smaller than that of a magnetometer. The peak signal magnitude of a tangential magnetometer decays with depth much faster than that of radial gradiometer, and the loss of the signal magnitude as a function of the dipole depth is fastest for planar gradiometers with 15-mm baseline. When the dipole position approaches the sphere center, the planar gradiometer signal is about 80% smaller than that of a magnetometer.

In the absence of noise and provided that the measured fields are not undersampled, arrays of different flux transformers are equivalent and it is possible to predict signals detected by one flux transformer array type from the measurement of another [133, 134]. However, in realistic measurement situations, the sensors are exposed not only to the wanted brain signals, but also to the environmental noise, noise from electrically active tissues other than brain, unwanted brain signals, and the SQUID and electronics noise. The presence of noise removes the equivalence among different flux transformer arrays, since information that is lost in the noise cannot be recovered. Thus the noiseless flux transformer behavior, as shown in Figure 5.10b, is not sufficient for the determination of the optimum flux transformer configuration. In the presence of noise, the most important parameter is the SNR, and the flux transformer arrays must be optimized for a maximum SNR. In addition to noise suppression, sensor arrays must also be designed to extract the maximum possible information from the measured signal.

5.5.4.2.1 *Radial and Vector Magnetometers*
Most MEG systems use radial gradiometers, magnetometers, planar gradiometers, or combinations of magnetometers and planar gradiometers. It has been suggested [102, 105, 135] that the vector magnetometers (or "vector" gradiometers) would produce more information because the tangential field components would also be measured.

For random SQUID noise and a single dipole, it was shown by Monte Carlo simulations, by expansion of the dipole equations in the vicinity of the correct dipole position, and by considering the information content that for a given number of channels it is best to deploy all channels as radial rather than vector devices [136, 137]. This is illustrated in Figure 5.11b, where the dipole localization error, σ_V, multiplied by $\sqrt{K}$, where K is the number of channels, is shown as a function of the dipole position relative to a hemispherical sensor shell [132]. Each value of σ_V was determined by 200 Monte Carlo simulations and the correlated noise was simulated using 1000 random magnitude and randomly distributed and oriented dipoles in the model sphere. The figure indicates that if the dipole is within the helmet or even slightly outside it, the radial magnetometers produce smaller localization error, independent of the number of channels used.

If the correlated brain noise and a single dipole are considered, as in Figure 5.11c, the conclusions are different. For dipoles well within the sensor shell, the localization errors for both radial and vector devices are similar. However, when the dipole is moved to the sensor shell rim or slightly outside it, the vector magnetometers produce smaller localization errors (vector magnetometers can extrapolate better in the vicinity of the sensor array edge). This conclusion was reached assuming that the centers of the vector and radial magnetometers are at the same distance from the head (point coils). Since this is physically not possible, the vector magnetometer centers must be moved

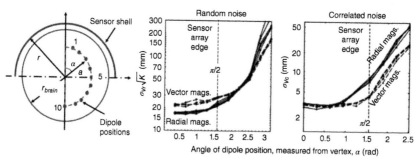

Figure 5.11 Comparison of radial and vector magnetometer dipole localization accuracy, σ_V, as a function of the ECD position relative to the sensor array edge, $r_{brain} = 80$ mm, hemispherical sensor shell, $r = 107$ mm, $a = 70$ mm, point magnetometers, dipole positions in 18° increments, position 5 corresponds to the dipole at the sensor array edge. Number of investigated magnetometers: radial $K = 97, 142, 190, 289, 586$; vector: $K = 96, 141, 291, 570$ (numbers of the vector magnetometer sites 32, 47, 97, 190). Lines in (b) and (c) correspond to constant number of channels; solid lines, radial magnetometers; dashed lines, vector magnetometers. Vertical dashed line marks the array edge. (a) Simulation geometry. (b) Random noise of 30 fT rms (5 fT/$\sqrt{Hz}$ in 36 Hz bandwidth). (c) Correlated noise simulated to resemble brain noise in 1-Hz bandwidth. Radial magnetometers: 30.4 fT rms; vector magnetometers: 22.5 fT rms (rms value was computed over all three vector components). The two different noise values correspond to the same random distribution of dipoles (used for the brain noise simulation). *Source:* adapted from [36].

slightly farther away from the head in order to accommodate the coil dimension (see Figure 5.10a). If the vector magnetometer centers were, for example, 10 mm farther from the head than the radial magnetometer centers, all advantage of the vector magnetometers in the presence of the correlated noise would be lost, that is, the vector and radial magnetometer performances for sources outside the helmet would be comparable, while for the sources within the helmet the vector magnetometer localization error would be larger.

It has been argued [135] that if multiple sources with overlapping waveforms are present, then the vector magnetometer information can result in better localization. However, the experiments were performed with a relatively small-area array (not a whole-cortex system) and no comparative analysis of the radial and vector devices with equal number of channels was performed.

It is sometimes argued that measurement of the radial field components is preferable because they are less dependent on the volume currents than the tangential components [138]. However, simulations with realistically shaped conductor models showed that the disturbances of the conductor shape produce equal distortions in all three field components [139]. This conclusion is to be expected because any field component can be computed from the distribution of any other component by integrating curl $B = 0$ in the current-free space outside the source region [122].

5.5.5 Magnetocardiography

Measurements of electric potential differences on the body surface that arise from the heart (electrocardiography [ECG]) have an important and established role in clinical diagnosis and research of the cardiac function. Both the 12-lead ECG and multilead body surface potential mapping (BSPM) are used [140]. The same bioelectric activity in the body that generates electric potentials also gives rise to biomagnetic fields, which are extremely low in magnitude, although much stronger than the MEG signals. The first successful recording of the MCG was reported in 1963 [8]. However, it was only after the development of SQUID sensors at the beginning of the 1970s that accurate detection of MCG signals became possible [9]. Until the 1990s, most MCG studies were performed with single-channel devices, by moving the system sequentially over the thorax and measuring signals from one location at time. Today, noninvasive MCG mapping recordings are carried out with multichannel systems, acquiring the signals simultaneously over the whole chest of the patient or at least a good part of it [100, 141, 142]. More information can thus be acquired than with the standard 12-lead ECG. Furthermore, vector-type gradiometers detecting all three field components were demonstrated, possibly offering additional advantages [135, 143, 144]. Efficient conversion methods from one sensor configuration to another facilitate accurate comparisons of data between different sensor arrays [134, 145, 146]. In addition, difficulties with skin–electrode contact are avoided, which sometimes causes problems in ECG studies. However, one should not consider MCG to be a possible replacement of ECG. The much higher equipment cost and the necessity of cooling to liquid helium temperatures would make it impractical. High-T_c SQUID systems operating in liquid nitrogen can somewhat reduce the cost of cooling, but at a price of reduced sensitivity [106, 108, 147].

Although MCG has not yet been established as a routine clinical tool, successful results have been reported in several clinically important problems, including assessment of the risk of life-threatening arrhythmias, detection and characterization of coronary diseases, noninvasive localization of cardiac activation, such as arrhythmia-causing regions or ischemic currents, and studies of fetal heart function. This section considers the basic principles of cardiomagnetism, focusing on MCG instrumentation and measurements.

5.5.5.1 Cardiomagnetic Instrumentation

Cardiomagnetic SQUID systems follow similar basic principles to the multichannel MEG sensors. The DC SQUID sensors offer the best sensitivity for the measurement of cardiomagnetic fields. Still, strong environmental magnetic noise, unavoidable at urban hospitals and laboratories, makes the detection of biomagnetic signals impossible without special techniques for environmental interference suppression. The environmental magnetic noise is reduced by MSR, which typically consist of a combination of μ-metal and eddy current shields. In addition, gradiometer coils are used to diminish residual magnetic noise within the shields. Alternatively, high-order gradiometers can be utilized if no magnetic shielding is employed.

Besides wire-wound gradiometer coils, arrays of magnetometer sensors can be utilized to compose differences between one or several coils electronically [100, 148]. Alternatively, software-based interference suppression methods can be applied, such as synthetic gradiometers or signal space projection, noise reduction with wavelets [149], or the signal space separation technique [150].

Conventionally, sensors in MCG instruments are positioned in one plane (or nearly so) to detect the magnetic field component perpendicular to the chest surface of a subject in the supine position. This field component, usually denoted as B_z, is analogous to the radial field component in MEG studies. Several MCG instruments have utilized first-order axial gradiometers with relatively large coils [100, 101, 141, 148, 151, 152]. However, planar magnetometers and gradiometers on integrated thin-film chips use less space and are easier to manufacture [153, 154]. Their small pickup area and the short baseline of gradiometers are not a problem if a sufficient signal-to-noise (SNR) reserve exists, as is the case in a good-quality MSR. Sensor configurations measuring all three field components have also been constructed [135, 143, 144].

Until the 1990s, biomagnetic measurements were usually performed with single or at most few-channel instruments cooled by liquid helium [141]. However, most MCG applications require mapping over multiple locations of the thorax surface. With only one channel this is time-consuming and cannot provide simultaneous recordings, which are preferable or necessary in most clinical applications. Therefore, in the past decade, true multichannel systems were developed. Furthermore, the sensitivity of sensors has been improved. Typical noise levels for magnetometers or axial gradiometers range between 5 and $10\,fT/\sqrt{Hz}$ at low frequencies inside a MSR [101, 141, 142, 148, 151–153] and a sensitivity as low as $2\,fT/\sqrt{Hz}$ has been reported [100, 144].

Large-array multichannel MCG systems may contain more than 60 SQUID sensors in an array covering an area of up to 0.3 m in diameter over the subject's chest. The sensors are placed at the bottom part of a large cylindrical dewar with a flat or slightly curved bottom to fit the chest surface. The boil-off rate of helium in such a dewar is typically a few liters per day. The dewar is attached to a gantry system, which allows easy positioning of the sensor above the subject's thorax. It is often also possible to move the subject bed horizontally and vertically.

The position of the sensor array with respect to the subject's thorax can be determined, for example, by using special marker coils attached to the skin. The positions of these coils are determined by a three-dimensional digitizer before the measurement (in torso coordinates), and from the MCG recordings (in device coordinates) when electric current is fed to the coils. The MCG recordings are usually triggered by ECG (using nonmagnetic electrodes). For this purpose, a three-lead ECG is sufficient. However, in research studies, the measurements are often accompanied by 12-lead ECG or 32- to 128-channel BSPM recordings with the same or separate data acquisition as in MCG. Furthermore, nonmagnetic exercise ergometers have been developed to perform physical stress-testing in the supine position [142, 151, 154].

5.5.6 Magnetoneurography

5.5.6.1 History of Measuring Signal Propagation in Nerves

In 1850, Hermann Helmholtz described an experiment that enabled him to determine the velocity of impulse propagation along a frog's nerve fiber to be 27 m/s. This study of a peripheral nerve may be considered the beginning of quantitative experimental electrophysiology, which later included study of the brain and the heart. Biomagnetic research focused first on the stronger signals of the heart and the brain and it took a number of attempts by several investigators before Wikswo et al. finally succeeded in measuring the magnetic field generated by a frog's sciatic nerve [155]. In this experiment, the isolated nerve was kept at room temperature in a saline solution and was threaded through a ferrite ring. The toroidal pickup coil wound around this ring was coupled to a helium-cooled SQUID. This design concentrated the nerve's magnetic flux in the ferrite ring and its signal was multiplied by the toroidal coil. Although at first sight this approach appeared to be restricted to in vitro or ex vivo preparations, a modification was developed that allowed this technique to be applied in vivo and in humans. To this end, an openable clip-on toroid was constructed that could be closed around a human nerve exposed in a saline solution during a surgical intervention [156]. In this arrangement, helium-cooled devices could be kept at a distance from the investigated subject. This concept was developed even further and resulted in the low-cost biomagnetic current probe system, an alternative recording technique that did not even need a SQUID.

The first noninvasive in vivo recordings of the magnetic field of action currents of the human median nerve were reported a few years after this pioneering work [157–160]. These studies were designed along the established paradigm for conventional neuroelectric recordings: a peripheral nerve in the arm of a human subject was stimulated at the wrist by an electric current impulse of about 15 mA amplitude and 100 µs duration mediated by a bipolar electrode that was decoupled from the ground to prevent power line interference. The nerve's signal was recorded as it propagated toward the brain by a single- or multichannel SQUID device in a conventional dewar operated in a MSR. When it was positioned above the upper arm of a healthy subject, a deflection of about 100 fT amplitude and about 1–2 ms duration was recorded a few milliseconds after the stimulation pulse.

Studies on other nerves followed, in particular on the ulnar nerve in the arm and the tibial nerve in the leg [161, 162]. Different regions along the propagation path of these nerves were measured, such as the lumbar region for the tibial nerve and the plexus in the shoulder where the arm nerves divide into several branches [163–165]. Further investigations were on the cervical region where the particular features of the neck as a volume conductor make straightforward interpretation of the data difficult [162, 166]. It became evident during early investigations (and remains still valid today) that:

1) MNG signals are among the weakest biomagnetic signals studied so far. Their amplitude is only about 5–10 fT in the lumbar region, which is the area of most interest for diagnostic applications. This requires high sensitivity measurements and sophisticated signal processing technology.
2) Because the generators in peripheral nerves have a relatively simple structure, peripheral nerve studies may serve as a test bed for biomagnetic measurement techniques and biomagnetic modeling.
3) The simple source structure makes it relatively easy to find the corresponding physical models and to extract a number of diagnostically relevant features of peripheral nerve function from MNG data.
4) MNG allows one to proceed from the level of the signal to the level of generators. Functional information in terms of current distributions inside the nerve tissue, as obtained with noninvasive MNG, complements the surface information measured by conventional electroneurography (ENG).

5.5.6.2   Measurement Technique and Signal Processing

MNG fields have to be recorded with a bandwidth of several kilohertz, because of their short time constants. This implies that a large broadband noise is added to the MNG data and reduces their SNR. An obvious, but important approach to noise reduction is the use of high-performance recording devices for MNG, that is, low-noise SQUIDs including low-noise electronics, and good shielding. Fortunately, MNG is not too affected by low-frequency noise because the MNG signals usually do not extend below 20 Hz. In fact, it was shown that an eddy current shielding provided by a room made of one-inch-thick aluminum may attenuate the environmental noise sufficiently for recording MNG [167]. It is advisable to use low-T_c SQUIDs for MNG, but it illustrates the good performance of modern high-T_c SQUIDs that they also can meet the requirements for MNG [168]. However, the best possible field sensitivity of low-T_c devices is clearly desirable for MNG.

Noise can be reduced by extensive averaging. Since impulse propagation in peripheral nerves is known to be stable and reproducible, there is no technical or physiological limit to this approach. Practical experience has shown, however, that even healthy volunteers are hardly able to undergo this uncomfortable and stressful investigation in a relaxed condition for much more than 15 minutes. With a stimulus repetition rate of about 10 Hz this limits the recording capacity to about 10 000 events. Assuming an instrumental noise of $3 \, \text{fT}/\sqrt{\text{Hz}}$ and a recording bandwidth of 2.5 kHz, this results in an rms value of noise of around 1.5 fT. This does not look too bad for analyzing MNG signals of 5–10 fT amplitude, but, unfortunately, there is additional noise originating from physiological sources. Even if patients are cooperative and remain motionless with relaxed muscles, the remaining muscle activities produce strong unstable and nonstationary perturbations in the MNG frequency range of interest. In addition, the periodic perturbation due to heart signals usually has an amplitude that is orders of magnitude larger than the MNG. The presence of these perturbations of physiological sources requires special signal processing tools. To this end, the concept of weighted averaging has proved to be most useful [169]. The main idea of weighted averaging is to weight each epoch in the averaging procedure by its signal power. This method reduces the impact of noisy samples in the averaging procedure, but it also results in an attenuation of the signal. The unwanted effect of signal underestimation can be compensated by an iterative procedure [170]. To suppress the periodical interference by heart signals in the triggered recordings, it is also helpful to include the noise covariance in the weighted averaging [171]. All together, these signal processing tools have provided field distributions with a sufficient SNR to allow an analysis of MNG signals in terms of dedicated mathematical and physical models.

5.5.6.3 Source Modeling for Magnetoneurography

Approximately 1000 to 10 000 axons are bundled to make up a peripheral nerve, which may branch toward the periphery, for example, to the fingers, as well as toward the spine, where three or four branches enter the spinal cord through different intervertebral root entrances. Individual axons have different conduction velocities, which, after a few tens of millimeters of propagation, may lead to a dispersion of the so-called compound action current, that is, the net current of the nerve bundle. The corresponding superposition of depolarization and repolarization currents may lead to partial cancellation of the net current and blurring of the corresponding neuromagnetic field, which can no longer be ascribed to a single current dipole. For the analysis of these fields, the current multipole expansion is a useful concept. It is well understood from theoretical considerations that different multipole components display different structural properties of the source. The ECD corresponds to the integral current intensity. A typical human peripheral nerve, like the tibial nerve, exhibits a current dipole moment of about 10 nA m. Note that this value is derived from MNG studies [162]. A measurement of the electric potential cannot provide this information directly, because body tissue conductivities must also be known for the current strength to be calculated. The quadrupolar term represents the inhomogeneity of the current distribution. For a linearly extended nerve, such inhomogeneity corresponds to two nearby nerve regions carrying currents in opposing directions (such as the currents corresponding to the depolarization and repolarization of the transmembrane potential). It was shown experimentally that these quadrupolar features are present in MNG recordings [172]. In particular, micro-SQUID MNG at the wrist, where the median nerve is close to the skin surface and the distance between the pickup coil and the nerve bundle is small, provided detailed spatial resolution of the compound action field structure [173]. Higher multipole expansion terms are of even more practical significance for gaining diagnostically relevant information from MNG. One of the octopole components represents the linear extent of a current distribution, which is important for understanding the effects of dispersion of the net current in the nerve bundle. By estimating this component from MNG data, the linear extent of the compound action current of the tibial nerve in the lumbar region, that is, after approximately 1 m propagation through the leg, was determined to be 140 mm [162], which is in agreement with physiological expectations.

5.5.6.4 Clinical Perspective

Peripheral nervous system diseases may be manifested as disturbances of the conduction velocity, distributed homogeneously or inhomogeneously along the conduction path. For these diseases, MNG is a diagnostic tool for which there is no counterpart in conventional electrophysiology. It was shown recently that MNG allows a much better reconstruction of the propagation path in stimulated nerves than multichannel ENG [174]. In terms of a current distribution model reconstructed from the MNG, better assessment of the affected function, that is the instantaneous conduction velocity, is possible. It would be worthwhile to find out whether this methodological advantage of MNG can provide additional diagnostic value for the treatment of diseases such as Guillain–Barre syndrome. A frequent disturbance in the peripheral nerve function is a conduction block due to a lesion, for instance, a nerve root compression. For a considerable percentage of affected patients, conventional imaging techniques display the structural lesions, but do not provide an unambiguous picture of the functional defect location. In such cases, MNG may provide additional information on the functional disturbance [175]. Magnetoneurography is a powerful tool for diagnosing localized malfunctions of the peripheral nervous system. Considering the frequency of such diseases in the population of industrialized countries, one may envision its high clinical potential. However, MNG requires the use of sophisticated data processing techniques, as well as a multichannel device having a flat array of SQUIDs with a noise level well below 5 fT/$\sqrt{\text{Hz}}$. Such instruments are not widely distributed so that these requirements may delay the application of MNG as a routine clinical diagnostic tool.

References

1 Heidari, H., Bonizzoni, E., Gatti, U., and Maloberti, F. (2015). A CMOS current-mode magnetic Hall sensor with integrated front-end. *IEEE Transactions on Circuits and Systems I: Regular Papers* 62 (5): 1270–1278.

2 Heidari, H., Bonizzoni, E., Gatti, U. et al. (2016). CMOS vertical Hall magnetic sensors on flexible substrate. *IEEE Sensors Journal* 16 (24): 8736–8743.

3 Nabaei, V., Chandrawati, R., and Heidari, H. (2018). Magnetic biosensors: modelling and simulation. *Biosensors and Bioelectronics* 103: 69–86.

4 Li, H., Shrestha, A., Heidari, H. et al. (2019). Magnetic and radar sensing for multimodal remote health monitoring. *IEEE Sensors Journal* 19 (20): 8979–8989.

5 Heidari, H., Bonizzoni, E., Gatti, U. et al. (2015). Optimal geometry of CMOS voltage-mode and current-mode vertical magnetic Hall sensors. In: *2015 IEEE SENSORS* (1 November 2015), 1–4. IEEE.

6 Manzin, A., Nabaei, V., and Ferrero, R. (2018). Quantification of magnetic nanobeads with micrometer Hall sensors. *IEEE Sensors Journal* 18 (24): 10058–10065.

7 Seidel, P. (2015). *Applied Superconductivity: Handbook on Devices and Applications*. Wiley.

8 Baule, G.M. (1963). Detection of the magnetic field of the heart. *American Heart Journal* 66: 95–96.

9 Cohen, D. (1970). Large-volume conventional magnetic shields. *Revue de Physique Appliquee* 5 (1): 53–58.

10 Zimmerman, J.E., Thiene, P., and Harding, J.T. (1970). Design and operation of stable rf-biased superconducting point-contact quantum devices, and a note on the properties of perfectly clean metal contacts. *Journal of Applied Physics* 41 (4): 1572–1580.

11 Ilmoniemi, R., Hari, R., and Reinikainen, K. (1984). A four-channel SQUID magnetometer for brain research. *Electroencephalography and Clinical Neurophysiology* 58 (5): 467–473.

12 Knuutila, J., Ahlfors, S., Ahonen, A. et al. (1987). Large-area low-noise seven-channel DC SQUID magnetometer for brain research. *Review of Scientific Instruments* 58 (11): 2145–2156.

13 Kajola, M. Ahlfors, S., Ehnholm, G.J. et al. (1989). A 24-channel magnetometer for brain research. In: *Advances in Biomagnetism* (ed. S.J. Williamson and M. Hoke), 673–676. Boston, MA: Springer.

14 Ahonen, A.I., Hamalainen, M.S., Kajola, M.J. et al. (1991). Multichannel SQUID systems for brain research. *IEEE Transactions on Magnetics* 27 (2): 2786–2792.

15 Kelhä, O.V., Erné, S.N., Hahlbohm, H.D., and Lübbig, H. (1981). *Biomagnetism*. Berlin: Walter de Gruyter.

16 Hoenig, H.E., Daalmans, G., Folberth, W. et al. (1989). Biomagnetic multichannel system with integrated SQUIDs and first order gradiometers operating in a shielded room. *Cryogenics* 29 (8): 809–813.

17 Hoenig, H.E., Daalmans, G.M., Bar, L. et al. (1991). Multichannel DC SQUID sensor array for biomagnetic applications. *IEEE Transactions on Magnetics* 27 (2): 2777–2785.

18 Becker, W., Diekmann, V., Jurgens, R., and Kornhuber, C. (1993). First experiences with a multichannel software gradiometer recording normal and tangential components of MEG. *Physiological Measurement* 14 (4A): A45.

19 Diekmann, V., Jürgens, R., Becker, W. et al. (1996). RF-SQUID to DC-SQUID upgrade of a 28-channel magnetoencephalography (MEG) system. *Measurement Science and Technology* 7 (5): 844.

20 Ripka, P. (2001). *Magnetic Sensors and Magnetometers*. Artech House.

21 Fagaly, R. (2006). Superconducting quantum interference device instruments and applications. *Review of Scientific Instruments* 77 (10): 101101.

22 London, F. (1954). *Superfluids*. Wiley.

23 Deaver, B.S. Jr. and Fairbank, W.M. (1961). Experimental evidence for quantized flux in superconducting cylinders. *Physical Review Letters* 7 (2): 43.

24 Doll, R. and Näbauer, M. (1961). Experimental proof of magnetic flux quantization in a superconducting ring. *Physical Review Letters* 7 (2): 51.

25 Clarke, J. and Braginski, A.I. (2004). *The SQUID Handbook: Fundamentals and Technology of SQUIDs and SQUID Systems*, vol. I. Weinheim: Wiley-VCH.

26 Grosz, A., Haji-Sheikh, M.J., and Mukhopadhyay, S.C. (2017). *High Sensitivity Magnetometers*. Springer.

27 Josephson, B.D. (1962). Possible new effects in superconductive tunnelling. *Physics Letters* 1 (7): 251–253.

28 McCumber, D.E. (1968). Effect of AC impedance on DC voltage-current characteristics of superconductor weak-link junctions. *Journal of Applied Physics* 39 (7): 3113–3118.

29 Stewart, W.C. (1968). Current-voltage characteristics of Josephson junctions. *Applied Physics Letters* 12 (8): 277–280.

30 Falco, C.M., Parker, W.H., Trullinger, S.E., and Hansma, P.K. (1974). Effect of thermal noise on current-voltage characteristics of Josephson junctions. *Physical Review B* 10 (5): 1865.

31 Voss, R.F. (1981). Noise characteristics of an ideal shunted Josephson junction. *Journal of Low Temperature Physics* 42 (1–2): 151–163.

32 Tesche, C.D. and Clarke, J. (1977). DC SQUID: noise and optimization. *Journal of Low Temperature Physics* 29 (3–4): 301–331.

33 Drung, D., Cantor, R., Peters, M. et al. (1990). Low-noise high-speed DC superconducting quantum interference device magnetometer with simplified feedback electronics. *Applied Physics Letters* 57 (4): 406–408.

34 Ketchen, M.B., Gallagher, W.J., Kleinsasser, A.W. et al. (1985). DC SQUID flux focuser. In: *SQUID'85: Superconducting Quantum Interference Devices and Their Applications.: Proceedings of the Third International Conference on Superconducting Quantum Devices*, Berlin (West) (25–28 June 1985) (ed. H.D. Hahlbohm and H. Lübbig) Walter de Gruyter.

35 Kirtley, J.R. (2010). Fundamental studies of superconductors using scanning magnetic imaging. *Reports on Progress in Physics* 73 (12): 126501.

36 Clarke, J. and Braginski, A.I. (2006). *The SQUID Handbook: Applications of SQUIDs and SQUID Systems*. Wiley.

37 Clem, T.R., Foley, C.P., and Keene, M.N. (2006). SQUIDs for geophysical survey and magnetic anomaly detection. In: *The SQUID Handbook: Applications of SQUIDs and SQUID Systems*, Vol. 2 (ed. J. Clarke and A.I. Braginski), Vol. 2 (ed. J. Clarke and A.I. Braginski), 481–543. Wiley.

38 Kraus, R. Jr., Espy, M., Magnelind, P., and Volegov, P. (2014). *Ultra-Low Field Nuclear Magnetic Resonance: A New MRI Regime*. Oxford University Press.

39 Jaycox, J. and Ketchen, M. (1981). Planar coupling scheme for ultra low noise DC SQUIDs. *IEEE Transactions on Magnetics* 17 (1): 400–403.

40 Ketchen, M. (1987). Integrated thin-film DC SQUID sensors. *IEEE Transactions on Magnetics* 23 (2): 1650–1657.

41 Knuutila, J., Kajola, M., Seppä, H. et al. (1988). Design, optimization, and construction of a DC SQUID with complete flux transformer circuits. *Journal of Low Temperature Physics* 71 (5–6): 369–392.

42 Cantor, R. (1996). DC SQUIDs: design, optimization and practical applications. In: *SQUID Sensors: Fundamentals, Fabrication and Applications* (ed. H. Weinstock), 179–233. Springer.

43 Zimmerman, J.E. (1971). Sensitivity enhancement of superconducting quantum interference devices through the use of fractional-turn loops. *Journal of Applied Physics* 42 (11): 4483–4487.

44 Dettmann, F., Richter, W., Albrecht, G., and Zahn, W. (1979). A monolithic thin film DC-SQUID. *Physica Status Solidi A* 51 (2): K185–K188.

45 Carelli, P. and Foglietti, V. (1982). Behavior of a multiloop DC superconducting quantum interference device. *Journal of Applied Physics* 53 (11): 7592–7598.

46 Drung, D., Knappe, S., and Koch, H. (1995). Theory for the multiloop DC superconducting quantum interference device magnetometer and experimental verification. *Journal of Applied Physics* 77 (8): 4088–4098.

47 Zakosarenko, V., Warzemann, L., Schambach, J. et al. (1996). Integrated LTS gradiometer SQUID systems for unshielded measurements in a disturbed environment. *Superconductor Science and Technology* 9 (4A): A112.

48 Jackel, L.D. and Buhrman, R.A. (1975). Noise in the rf SQUID. *Journal of Low Temperature Physics* 19 (3–4): 201–246.

49 Ehnholm, G.J. (1977). Theory of the signal transfer and noise properties of the rf SQUID. *Journal of Low Temperature Physics* 29 (1–2): 1–27.

50 Likharev, K.K. (1986). *Dynamics of Josephson Junctions and Circuits*. Gordon and Breach Science Publishers.

51 Ryhänen, T., Seppä, H., Ilmoniemi, R., and Knuutila, J. (1989). SQUID magnetometers for low-frequency applications. *Journal of Low Temperature Physics* 76 (5–6): 287–386.

52 Chesca, B. (1998). Theory of RF SQUIDS operating in the presence of large thermal fluctuations. *Journal of Low Temperature Physics* 110 (5–6): 963–1001.

53 Zeng, X.H., Zhang, Y., Chesca, B. et al. (2000). Experimental study of amplitude–frequency characteristics of high-transition-temperature radio frequency superconducting quantum interference devices. *Journal of Applied Physics* 88 (11): 6781–6787.

54 Kurkijärvi, J. (1973). Noise in the superconducting quantum flux detector. *Journal of Applied Physics* 44 (8): 3729–3733.

55 Hansma, P.K. (1973). Superconducting single-junction interferometers with small critical currents. *Journal of Applied Physics* 44 (9): 4191–4194.

56 Rifkin, R., Vincent, D.A., Deaver, B.S. Jr., and Hansma, P.K. (1976). rf SQUID's in the nonhysteretic mode: detailed comparison of theory and experiment. *Journal of Applied Physics* 47 (6): 2645–2650.

57 Kuzmin, L.S., Likharev, K.K., Migulin, V.V. et al. (1985). X-band parametric amplifier and microwave SQUID using single-tunnel-junction superconducting interferometer. In: *SQUID'85: Superconducting Quantum Interference Devices and Their Applications: Proceedings of the Third International Conference on Superconducting Quantum Devices*, Berlin (West) (25–28 June 1985) (ed. H.D. Hahlbohm and H. Lübbig). Walter de Gruyter.

58 Zhang, Y. (2001). Evolution of HTS rf SQUIDs. *IEEE Transactions on Applied Superconductivity* 11 (1): 1038–1042.

59 Drung, D. (1996). Advanced SQUID read-out electronics. In: *SQUID Sensors: Fundamentals, Fabrication and Applications* (ed. H. Weinstock), 63–116. Springer.

60 Koch, R.H., Clarke, J., Goubau, W.M. et al. (1983). Flicker (1/f) noise in tunnel junction DC SQUIDs. *Journal of Low Temperature Physics* 51 (1–2): 207–224.

61 Drung, D. (1995). Low-frequency noise in low-Tc multiloop magnetometers with additional positive feedback. *Applied Physics Letters* 67 (10): 1474–1476.

62 Oukhanski, N., Stolz, R., and Meyer, H.-G. (2006). High slew rate, ultrastable direct-coupled readout for DC superconducting quantum interference devices. *Applied Physics Letters* 89 (6): 063502.

63 Drung, D., Hinnrichs, C., and Barthelmess, H.-J. (2006). Low-noise ultra-high-speed DC SQUID readout electronics. *Superconductor Science and Technology* 19 (5): S235.

64 Cantor, R. and Ludwig, F. (2004). SQUID fabrication technology. In: *The SQUID Handbook: Fundamentals and Technology of SQUIDs and SQUID Systems*, vol. 1 (ed. J. Clarke and A.I. Braginski), 93–125. Weinheim: Wiley-VCH.

65 Hayakawa, H., Yoshikawa, N., Yorozu, S., and Fujimaki, A. (2004). Superconducting digital electronics. *Proceedings of the IEEE* 92 (10): 1549–1563.

66 Likharev, K.K. (2012). Superconductor digital electronics. *Physica C: Superconductivity and Its Applications* 482: 6–18.

67 Gates, J.V., Washington, M.A., and Gurvitch, M. (1984). Critical current uniformity and stability of Nb/Al-oxide-Nb Josephson junctions. *Journal of Applied Physics* 55 (5): 1419–1421.

68 Lehnert, T., Billon, D., Grassl, C., and Gundlach, K.H. (1992). Thermal annealing properties of Nb-Al/AlO x-Nb tunnel junctions. *Journal of Applied Physics* 72 (7): 3165–3168.

69 Kroger, H., Smith, L.N., and Jillie, D.W. (1981). Selective niobium anodization process for fabricating Josephson tunnel junctions. *Applied Physics Letters* 39 (3): 280–282.

70 Gurvitch, M., Washington, M.A., and Huggins, H.A. (1983). High quality refractory Josephson tunnel junctions utilizing thin aluminum layers. *Applied Physics Letters* 42 (5): 472–474.

71 Maezawa, M., Aoyagi, M., Nakagawa, H. et al. (1995). Specific capacitance of Nb/AlO x/Nb Josephson junctions with critical current densities in the range of 0.1–18 kA/cm^2. *Applied Physics Letters* 66 (16): 2134–2136.

72 Anders, S., Schmelz, M., Fritzsch, L. et al. (2009). Sub-micrometer-sized, cross-type Nb–AlOx–Nb tunnel junctions with low parasitic capacitance. *Superconductor Science and Technology* 22 (6): 064012.

73 Schmelz, M., Stolz, R., Zakosarenko, V. et al. (2010). SQUIDs based on submicrometer-sized Josephson tunnel junctions fabricated in a cross-type technology. *Superconductor Science and Technology* 24 (1): 015005.

74 Williamson, S.J. and Kaufman, L. (1981). Biomagnetism. *Journal of Magnetism and Magnetic Materials* 22 (2): 129–201.

75 Dössel, O., David, B., Fuchs, M. et al. (1993). A 31-channel SQUID system for biomagnetic imaging. *Applied Superconductivity* 1 (10–12): 1813–1825.

76 Nowak, H. (2007). *Magnetism in Medicine: A Handbook*. Wiley.

77 Huonker, R., Nowak, H., Rzanny, R., and Rieke, K. (1996). Combined 3D neuromagnetic source imaging and MRI-scans. *Electroencephalography and Clinical Neurophysiology Supplement* 47: 439–447.

78 Leder, U., Haueisen, J., Huck, M., and Nowak, H. (1998). Non-invasive imaging of arrhythmogenic left-ventricular myocardium after infarction. *The Lancet* 352 (9143): 1825.

79 Drung, D., Abmann, C., Beyer, J. et al. (2007). Highly sensitive and easy-to-use SQUID sensors. *IEEE Transactions on Applied Superconductivity* 17 (2): 699–704.

80 Drung, D. (2010). Simplified analysis of direct SQUID readout schemes. *Superconductor Science and Technology* 23 (6): 065006.

81 Drung, D., Storm, J.-H., and Beyer, J. (2013). SQUID current sensor with differential output. *IEEE Transactions on Applied Superconductivity* 23 (3): 1100204–1100204.

82 Vrba, J., Nenonen, J., and Trahms, L. (2006). Biomagnetism. In: *The SQUID Handbook. Vol. II: Applications of SQUIDs and SQUID Systems* (ed. J. Clarke and A.I. Braginski), 269–389. Wiley.

83 Nowak, H., Leder, U., Goernig, M. et al. (2003). Multichannel-vectormagnetocardiography: a new biomedical engineering approach. *Biomedizinische Technik/Biomedical Engineering* 48 (s1): 368–369.

84 Aittoniemi, K., Karp, P.J., Katila, T. et al. (1978). On balancing superconducting gradiometric magnetometers. *Le Journal de Physique Colloques* 39 (C6): C6-1223–C6-1225.

85 Overweg, J.A. and Walter-Peters, M.J. (1978). The design of a system of adjustable superconducting plates for balancing a gradiometer. *Cryogenics* 18 (9): 529–534.

86 Nowak, H., Giessler, F., and Huonker, R. (1991). Multichannel magnetography in unshielded environments. *Clinical Physics and Physiological Measurement* 12 (B): 5.

87 Williamson, S.J., Kaufman, L., and Brenner, D. (1977). Biomagnetism. In: *Superconductor Applications: SQUIDS and Machines* (ed. B. Schwartz), 355–402. Springer.

88 Drung, D. and Koch, H. (1993). An electronic second-order gradiometer for biomagnetic applications in clinical shielded rooms. *IEEE Transactions on Applied Superconductivity* 3 (1): 2594–2597.

89 Matlashov, A., Zhuravlev, Y., Lipovich, A. et al. (1989). Electronic noise suppression in multichannel neuromagnetic system. In: *Advances in Biomagnetism* (ed. S.J. Williamson and M. Hoke), 725–728. Boston, MA: Springer.

90 Tavrin, Y., Zhang, Y., Wolf, W., and Braginski, A.I. (1994). A second-order SQUID gradiometer operating at 77 K. *Superconductor Science and Technology* 7 (5): 265.

91 Vrba, J., Haid, G., Lee, S. et al. (1991). Biomagnetometers for unshielded and well shielded environments. *Clinical Physics and Physiological Measurement* 12 (B): 81.

92 Robinson, S.E. (1989). Environmental noise cancellation for biomagnetic measurements. In: *Advances in Biomagnetism* (ed. S.J. Williamson and M. Hoke), 721–724. Boston, MA: Springer.

93 ter Brake, H.J.M., Fleuren, F.H., Ulfrnan, J.A., and Flokstra, J. (1986). Elimination of flux-transformer crosstalk in multichannel SQUID magnetometers. *Cryogenics* 26 (12): 667–670.

94 Vrba, J. et al. (1993). Whole cortex, 64 channel SQUID biomagnetometer system. *IEEE Transactions on Applied Superconductivity* 3 (1): 1878–1882.

95 Knuutila, J.E.T. et al. (1993). A 122-channel whole-cortex SQUID system for measuring the brain's magnetic fields. *IEEE Transactions on Magnetics* 29 (6): 3315–3320.

96 Hydrospan www.hydrospan.cn/en (accessed 16 August 2019).

97 AtB. www.atb-it.com (accessed 16 August 2019).

98 ter Brake, H.J.M., Flokstra, J., Jaszczuk, W. et al. (1991). The UT 19-channel DC SQUID based neuromagnetometer. *Clinical Physics and Physiological Measurement* 12 (B): 45.

99 Matlashov, A., Slobodchikov, V., Bakharev, A. et al. (1995). Biomagnetic multichannel system built with 19 cryogenic probes. *Biomagnetism: Fundamental Research and Clinical Applications*: 493–496.

100 Drung, D. (1995). The PTB 83-SQUID system for biomagnetic applications in a clinic. *IEEE Transactions on Applied Superconductivity* 5 (2): 2112–2117.

101 Dossel, O., David, B., Fuchs, M. et al. (1993). A modular 31-channel SQUID system for biomagnetic measurements. *IEEE Transactions on Applied Superconductivity* 3 (1): 1883–1886.

102 Yoshida, Y., Arakawa, A., Kondo, Y. et al. (2000). A 129-channel vector neuromagnetic imaging system. In: *Biomag 96, Volume 1/Volume 2 Proceedings of the Tenth International Conference on Biomagnetism* (ed. C.J. Aine, Y. Okada, G. Stroink, et al.), 154–157. New York: Springer.

103 Ueda, M., Kandori, A., Ogata, H. et al. (1995). Development of a biomagnetic measurement system for brain research. *IEEE Transactions on Applied Superconductivity* 5 (2): 2465–2469.

104 Fujimoto, S., Sata, K., Fukui, N. et al. (2000). A 32-channel MCG system cooled by a GM/JT cryocooler. In: *Biomag 96 Volume 1/Volume 2 Proceedings of the Tenth International Conference on Biomagnetism*, (ed. C.J. Aine, Y. Okada, G. Stroink, et al.) 43–46. New York: Springer.

105 Kotani, M., Uchikawa, Y., Kawakatsu, M. et al. (1997). A whole-head SQUID system for detecting vector components. *Applied Superconductivity* 5 (7–12): 399–403.

106 Itozaki, H., Tanaka, S., Toyoda, H. et al. (1996). A multi-channel high-SQUID system and its application. *Superconductor Science and Technology* 9 (4A): A38.

107 Wikswo, J.P. (1995). SQUID magnetometers for biomagnetism and nondestructive testing: important questions and initial answers. *IEEE Transactions on Applied Superconductivity* 5 (2): 74–120.

108 Barthelmess, H.J., Halverscheid, M., Schiefenhovel, B. et al. (2001). Low-noise biomagnetic measurements with a multichannel DC-SQUID system at 77 K. *IEEE Transactions on Applied Superconductivity* 11 (1): 657–660.

109 Pflieger, M.E., Simpson, G.V., Ahlfors, S.P., and Ilmoniemi, R.J. (2000). Superadditive information from simultaneous MEG/EEG data. In: *Biomag 96, Volume 1/Volume 2 Proceedings of the Tenth International Conference on Biomagnetism* (ed. C.J. Aine, Y. Okada, G. Stroink, et al.), 1154–1157. New York: Springer.

110 Cohen, D. and Cuffin, B.N. (1987). A method for combining MEG and EEG to determine the sources. *Physics in Medicine and Biology* 32 (1): 85.

111 Lauterbur, P.C. (1973). Image formation by induced local interactions: examples employing nuclear magnetic resonance. *Nature* 242: 190–191.

112 Hinshaw, W.S. and Lent, A.H. (1983). An introduction to NMR imaging: from the Bloch equation to the imaging equation. *Proceedings of the IEEE* 71 (3): 338–350.

113 Swenberg, C.E., Conklin, J.J., and Haddad, R.A. (1989). Imaging techniques in biology and medicine. *Applied Optics* 28: 3416.

114 Ter-Pogossian, M.M., Phelps, M.E., Hoffman, E.J., and Mullani, N.A. (1975). A positron-emission transaxial tomograph for nuclear imaging (PETT). *Radiology* 114 (1): 89–98.

115 Gilardi, M.C., Rizzo, G., Lucignani, G., and Fazio, F. (1996). Integrating competing technologies with MEG. In: *SQUID Sensors: Fundamentals, Fabrication and Applications* (ed. H. Weinstock), 491–516. Springer.

116 Knoll, G.F. (1983). Single-photon emission computed tomography. *Proceedings of the IEEE* 71 (3): 320–329.

117 Stehling, M.K., Turner, R., and Mansfield, P. (1991). Echo-planar imaging: magnetic resonance imaging in a fraction of a second. *Science* 254 (5028): 43–50.

118 Belliveau, J.W., Kennedy, D.N., McKinstry, R.C., et al. (1991). Functional mapping of the human visual cortex by magnetic resonance imaging. *Science* 254 (5032): 716–719.

119 Partridge, L.D. and Partridge, L.D. (1993). *The Nervous System: Its Function and Its Interaction with the World.* Cambridge, MA: MIT Press.

120 Wikswo, J.P. (1989). Biomagnetic sources and their models. In: *Advances in Biomagnetism* (ed. S.J. Williamson and M. Hoke), 1–18. Boston, MA: Springer.

121 Taccardi, B. (1983). Electrophysiology of excitable cells and tissues, with special consideration of the heart muscle. In: *Biomagnetism* (ed. S.J. Williamson, G.-L. Romani, L. Kaufman, and I. Modena), 141–171. Springer.

122 Hämäläinen, M., Hari, R., Ilmoniemi, R.J. et al. (1993). Magnetoencephalography – theory, instrumentation, and applications to noninvasive studies of the working human brain. *Reviews of Modern Physics* 65 (2): 413.

123 Carpenter, M.B. (1985). *Core Text of Neuroanatomy*. Williams & Wilkins.

124 Tripp, J.H. (1983). Physical concepts and mathematical models. In: *Biomagnetism* (ed. S.J. Williamson, G.-L. Romani, L. Kaufman, and I. Modena) 101–139. Springer.

125 Swinney, K.R. and Wikswo, J.P. Jr. (1980). A calculation of the magnetic field of a nerve action potential. *Biophysical Journal* 32 (2): 719.

126 Grynszpan, F. and Geselowitz, D.B. (1973). Model studies of the magnetocardiogram. *Biophysical Journal* 13 (9): 911–925.

127 Sarvas, J. (1987). Basic mathematical and electromagnetic concepts of the biomagnetic inverse problem. *Physics in Medicine & Biology* 32 (1): 11.

128 Hillebrand, A. and Barnes, G.R. (2002). A quantitative assessment of the sensitivity of whole-head MEG to activity in the adult human cortex. *Neuroimage* 16 (3): 638–650.

129 Nakasato, N., Fujita, S., Seki, K. et al. (1995). Functional localization of bilateral auditory cortices using an MRI-linked whole head magnetoencephalography (MEG) system. *Electroencephalography and Clinical Neurophysiology* 94 (3): 183–190.

130 Okada, Y. (2003). Toward understanding the physiological origins of neuromagnetic signals. In: *Magnetic Source Imaging of the Human Brain* (ed. Z.-L. Lu and L. Kaufman), 43–76. Mahwah, NJ: Lawrence Erlbaum Associates.

131 Fagaly, R.L. (1990). Neuromagnetic instrumentation. *Advances in Neurology* 54: 11–32.

132 Vrba, J. (2000). Multichannel SQUID biomagnetic systems. In: *Applications of Superconductivity* (ed. H. Weinstock), 61–138. Springer.

133 Ahonen, A.I., Hämäläinen, M.S., Ilmoniemi, R.J. et al. (1993). Sampling theory for neuromagnetic detector arrays. *IEEE Transactions on Biomedical Engineering* 40 (9): 859–869.

134 Ilmoniemi, R.J. (1992). Synthetic magnetometer channels for standard representation of data. In: *Biomagnetism: Clinical Aspects: Proceedings of the 8th International Conference on Biomagnetism*, Münster (19–24 August 1991) (ed. M. Hoke). Excerpta Medica.

135 Uchikawa, Y., Kobayashi, K., Kawakatsu, M., and Kotani, M. (2001). A 3-D vector measurement and its application on biomagnetic signals. *Proceedings of the 12th International Conference on Biomagnetism, BIOMAG2001*, Espoo, Finland (13–17 August 2001), vol. 915.

136 Hughett, P. and Miyauchi, S. (2000). A comparison of vector and radial magnetometer arrays for whole-head magnetoencephalography. In: *Biomag 96, Volume 1/Volume 2 Proceedings of the Tenth International Conference on Biomagnetism* (ed. C.J. Aine, Y. Okada, G. Stroink, et al.), 51–54. New York: Springer.

137 Matsuba, H., Vrba, J., and Cheung, T. (2000). Current dipole localization errors as a function of the system noise and the number of sensors. In: *Biomag 96, Volume 1/Volume 2 Proceedings of the Tenth International Conference on Biomagnetism* (ed. C.J. Aine, Y. Okada, G. Stroink, et al.), 79–82. New York: Springer.

138 Cuffin, B.N. and Cohen, D. (1977). Magnetic fields of a dipole in special volume conductor shapes. *IEEE Transactions on Biomedical Engineering* 24 (4): 372–381.

139 Hamalainen, M.S. and Sarvas, J. (1989). Realistic conductivity geometry model of the human head for interpretation of neuromagnetic data. *IEEE Transactions on Biomedical Engineering* 36 (2): 165–171.

140 MacFarlane, P.W. and Veitch Lawrie, T.D. (eds.) (1989). *Comprehensive Electrocardiology. Theory and Practice in Health and Disease.* New York: Pergamon Press.

141 Nowak, H. (2006). Biomagnetic instrumentation. In: *Magnetism in Medicine: A Handbook*, 2e (ed. W. Andrä and H. Nowak), 101–163. Weinheim: Wiley-VCH.

142 Erné, S.N., Pasquarelli, A., Kammrath, H. et al. (1998). Argos 55-the new MCG system in Ulm. In: *BIOMAG-98. Proceedings of the 11th International Conference on Biomagentism*, Sendai, Japan, pp. 27–30

143 Burghoff, M., Schleyerbach, H., Drung, D. et al. (1999). A vector magnetometer module for biomagnetic application. *IEEE Transactions on Applied Superconductivity* 9 (2): 4069–4072.

144 Schnabel, A., Burghoff, M., Hartwig, S. et al. (2004). A sensor configuration for a 304 SQUID vector magnetometer. *Neurology & Clinical Neurophysiology: NCN* 2004: 70–70.

145 Numminen, J., Ahlfors, S., Ilmoniemi, R. et al. (1995). Transformation of multichannel magnetocardiographic signals to standard grid form. *IEEE Transactions on Biomedical Engineering* 42 (1): 72–78.

146 Burghoff, M., Nenonen, J., Trahms, L., and Katila, T. (2000). Conversion of magnetocardiographic recordings between two different multichannel SQUID devices. *IEEE Transactions on Biomedical Engineering* 47 (7): 869–875.

147 Itozaki, H., Sakuta, K., Kobayashi, T. et al. (2003). Applications of HTSC SQUIDs. In: *Vortex Electronis and SQUIDs* (ed. T. Kobayashi, H. Hayakawa, and M. Tonouchi), 185–248. Berlin: Springer.

148 Van Leeuwen, P., Haupt, C., Hoormann, C. et al. (1999). A 67-channel biomagnetometer designed for cardiology and other applications. In: *Recent Advances in Biomagnetism* (ed. T. Yoshimoto, M. Kotani, S. Kuriki, and H. Karibe), 89–92. Sendai: Tohoku University Press.

149 Sternickel, K., Effern, A., Lehnertz, K. et al. (2001). Nonlinear noise reduction using reference data. *Physical Review E* 63 (3): 036209.

150 Taulu, S., Kajola, M., and Simola, J. (2004). Suppression of interference and artifacts by the signal space separation method. *Brain Topography* 16 (4): 269–275.

151 Schneider, S., Hoenig, E., Reichenberger, H. et al. (1990). Multichannel biomagnetic system for study of electrical activity in the brain and heart. *Radiology* 176 (3): 825–830.

152 Tsukada, K., Kandori, A., Miyashita, T. et al. (1998). A simplified superconducting quantum interference device system to analyze vector components of a cardiac magnetic field. In: *Proceedings of the 20th Annual International Conference of the IEEE Engineering in Medicine and Biology Society. Vol. 20 Biomedical Engineering Towards the Year 2000 and Beyond* (Cat. No. 98CH36286) (1 November 1998), 524–527. IEEE.

153 Montonen, J., Ahonen, A., Hämäläinen, M. et al. (2000). Magnetocardiographic functional imaging studies in the BioMag laboratory. In: *Biomag 96* (ed. C.J. Aine, Y. Okada, G. Stroink, et al.), 494–497. New York: Springer.

154 Montonen, J., Ahonen, A., Hämäläinen, M. et al. (2000). Magnetocardiographic functional imaging studies in BioMag Laboratory. In: *Advances in Biomagnetism Research: Biomag96* (ed. C. Aine, Y. Okada, G. Stroink, et al.). New York: Springer Verlag.

155 Wikswo, J.P., Barach, J.P., and Freeman, J.A. (1980). Magnetic field of a nerve impulse: first measurements. *Science* 208 (4439): 53–55.

156 Leifer, M.C. and Wikswo, J.P. Jr. (1983). Optimization of a clip-on SQUID current probe. *Review of Scientific Instruments* 54 (8): 1017–1022.

157 Erné, S.N., Curio, G., Trahms, L. et al. (1988). Magnetic activity of a single peripheral nerve in man. *Biomagnetism* 87: 166–169.

158 Trahms, L., Erné, S.N., Trontelj, Z. et al. (1989). Biomagnetic functional localization of a peripheral nerve in man. *Biophysical Journal* 55 (6): 1145–1153.

159 Hari, R., Hällström, J., Tiihonen, J., and Joutsiniemi, S.L. (1989). Multichannel detection of magnetic compound action fields of median and ulnar nerves. *Electroencephalography and Clinical Neurophysiology* 72 (3): 277–280.

160 Hashimoto, I., Odaka, K., Gatayama, T., and Yokoyama, S. (1991). Multichannel measurements of magnetic compound action fields of the median nerve in man. *Electroencephalography and Clinical Neurophysiology/Evoked Potentials Section* 81 (5): 332–336.

161 Hashimoto, I., Mashiko, T., Mizuta, T. et al. (1995). Multichannel detection of magnetic compound action fields with stimulation of the index and little fingers. *Electroencephalography and Clinical Neurophysiology/ Electromyography and Motor Control* 97 (2): 102–113.

162 Mackert, B.-M., Curio, G., Burghoff, M., and Marx, P. (1997). Mapping of tibial nerve evoked magnetic fields over the lower spine. *Electroencephalography and Clinical Neurophysiology/Evoked Potentials Section* 104 (4): 322–327.

163 Curio, G., Erné, S.N., Sandfort, J. et al. (1991). Exploratory mapping of evoked neuromagnetic activity from human peripheral nerve, brachial plexus and spinal cord. *Electroencephalography and Clinical Neurophysiology/Evoked Potentials Section* 81 (6): 450–453.

164 Curio, G., Reill, L., Sandfort, J. et al. (1993). Nerve, plexus and spinal cord: possible targets for non-invasive neuromagnetic measurements in man. *Physiological Measurement* 14 (4A): A91.

165 Mackert, B.-M., Burghoff, M., Hiss, L.-H. et al. (2000). Non-invasive magnetoneurography for 3D-monitoring of human compound action current propagation in deep brachial plexus. *Neuroscience Letters* 289 (1): 33–36.

166 Mackert, B.-M., Burghoff, M., Hiss, L.H. et al. (2001). Magnetoneurography of evoked compound action currents in human cervical nerve roots. *Clinical Neurophysiology* 112 (2): 330–335.

167 Lang, G., Shahani, U., Weir, A.I. et al. (1998). Neuromagnetic recordings of the human peripheral nerve with planar SQUID gradiometers. *Physics in Medicine & Biology* 43 (8): 2379.

168 Drung, D., Ludwig, F., Müller, W. et al. (1996). Integrated $YBa_2Cu_3O_{7-x}$ magnetometer for biomagnetic measurements. *Applied Physics Letters* 68 (10): 1421–1423.

169 Lütkenhöner, B., Hoke, M., and Pantev, C. (1985). Possibilities and limitations of weighted averaging. *Biological Cybernetics* 52 (6): 409–416.

170 Burghoff, M., Mackert, B.M., Haberkorn, W. et al. (1999). High-resolution magnetoneurography. *Applied Superconductivity* 6 (10–12): 567–575.

171 Gräbe, T., Burghoff, M., Steinhoff, U. et al. (1997). Identification of signal components by stochastic modelling in measurements of evoked magnetic fields from peripheral nerves. In: *Series on Advances in Mathematics for Applied Sciences: Volume 45: Advanced Mathematical Tools in Metrology III* (ed. P. Ciarlini, M.G. Cox, F. Pavese, and D. Richter), 178–185.

172 Hashimoto, I., Mashiko, T., Mizuta, T. et al. (1994). Visualization of a moving quadrupole with magnetic measurements of peripheral nerve action fields. *Electroencephalography and Clinical Neurophysiology/Evoked Potentials Section* 93 (6): 459–467.

173 Hoshiyama, M., Kakigi, R., and Nagata, O. (1999). Peripheral nerve conduction recorded by a micro gradiometer system (micro-SQUID) in humans. *Neuroscience Letters* 272 (3): 199–202.

174 Mackert, B.-M., Burghoff, M., Hiss, L.-H. et al. (2001). Tracing of proximal lumbosacral nerve conduction – a comparison of simultaneous magneto- and electroneurography. *Clinical Neurophysiology* 112 (8): 1408–1413.

175 Mackert, B.-M., Curio, G., Burghoff, M. et al. (1998). Magnetoneurographic 3D localization of conduction blocks in patients with unilateral S1 root compression. *Electroencephalography and Clinical Neurophysiology/ Electromyography and Motor Control* 109 (4): 315–320.

6

Conclusion

6.1 Outlook

Magnetic field detection has vastly expanded as industry has utilized a variety of magnetic sensors to detect the presence, strength, or direction of magnetic fields not only from the Earth but also from permanent magnets, magnetized soft magnets, and the magnetic fields associated with current. These sensors are used as proximity sensors, speed and distance measuring devices, navigation compasses, and current sensors. They can measure these properties without actual contact to the medium being measured and become the eyes of many control systems. Most of the common magnetic sensing methods have been described and the underlying physical principles governing their operation have been illustrated in this book. A varied set of applications that exploit specific characteristics of these sensors are also described.

The future trends in magnetic sensors should be discussed from these same two perspectives, physics and applications. In the past, discoveries of new physics phenomenon have led to new sensor technologies. Many of the phenomena exploited by sensors were discovered in the 1800s and early 1900s (i.e. the Faraday effect, the Hall effect, superconductivity, etc.). However, as discussed here, there have been several more recent discoveries that have affected magnetic sensor technology. For example, Josephson tunneling in superconducting structures was observed in the 1960s. More recently, giant magnetoresistance and magnetic tunnel junctions have significantly affected information storage and sensor technology. It is likely that future discoveries will open new possibilities for improved magnetic sensors [1]. From an applications perspective, the need for improved sensors is ubiquitous. Magnetic sensors are used when other sensors have unwanted signals from the changing environment. The trend is constantly toward smaller size, lower power consumption, and lower cost for similar or improved performance [2–7]. There is not much need to improve the sensitivity independent of size, power, and cost. Instead, for each

Magnetic Sensors for Biomedical Applications, First Edition. Hadi Heidari and Vahid Nabaei.
© 2020 by The Institute of Electrical and Electronics Engineers, Inc.
Published 2020 by John Wiley & Sons, Inc.

application, one needs to make a trade-off between sensitivity, size, power, and cost. The possible routes for enhancing the performance of magnetic sensors are (i) new phenomenon, (ii) new applications of existing phenomenon, (iii) improved materials, and (iv) improved processing and manufacturing. A major need is to reduce the cost of the signal processing electronics since, in many cases, the signal processing electronics is much more expensive than the sensor element.

Obtaining better material properties will be one of the major methods for making improvements in the size, power, and cost in magnetic sensors. Improvements in material properties are a result of the thousands of research studies published each year. It is hard to predict which materials will have the most significant improvement, and, most likely, in many cases, these improvements will come in minor steps.

There are, however, areas where large improvements are possible. For example, there has been research on half metals [8]. These are metals that have no minority spins states at the Fermi level. If MTJ sensors could be fabricated with materials that are half metals, the magnetoresistance would be infinite. It turns out to be difficult or impossible to produce materials that are half metals at room temperature. It is likely that the new field of spintronics will lead to improved magnetic sensors.

The appeal of co-integrating semiconductor devices, ferromagnetic pieces, and coils on the same chip is great. This would allow the development of magnetic sensor microsystems with drastically reduced offset and noise (thanks to the use of magnetic chopping), high sensitivity (magnetic concentration), high stability (autocalibration), and so on [9]. Due to such a high importance, research into the integration of ferromagnetic structures and coils will he pursued in the future with high priority. We expect that ways will be found to integrate both high-permeability and magnetically highly nonlinear ferromagnetic materials easily on a silicon wafer. Also the simplified structure and technology will be found to allow a co-integration of coils. Once available, low-cost one-chip magnetic sensor microsystems will revolutionize the use of magnetic sensors. Many new applications for magnetic sensors would become apparent, guaranteeing an exciting future for the corresponding microsystem research and development.

6.2 A Conclusion on Galvanomagnetic Sensors

Unlike other transducers of physical and chemical quantities, magnetic sensors have the unique ability to reveal realities that cannot be perceived by the human senses. Hence our knowledge of magnetic fields comes solely from these devices, combining in themselves different laws of Nature. After having seen semiconductor-based Hall effect devices in this book, one may ask: what is the relative importance of each of these devices? How do these devices compare with other galvanomagnetic sensors, such as thin-film ferromagnetic magnetoresistors? And how might this picture look in the years to come? We shall compare only magnetic field sensors here, which are similar to Hall sensors: that means solid-state magnetic sensors, operating around room temperature, low-cost, and suitable for many industrial and laboratory applications. Only some galvanomagnetic sensors fit these criteria. In order to make the comparison of various galvanomagnetic devices that meet the above criteria, the impact of these sensors both in scientific research and in industry can be investigated. What determines the impact of a magnetic sensor is what our peers in other research fields and in industry think about the potential of this device to solve their magnetic sensing problems.

6.2.1 Hall Elements: Hall Voltage Mode Versus Hall Current Mode and Magnetoresistance Mode of Operation

For an AC magnetic signal above the $1/f$-noise region, the signal-to-noise ratios of equivalent Hall plates working in the Hall voltage mode, in the Hall current mode, and in the magnetoresistance mode are similar [10]. However, for a DC and low-frequency magnetic signal, this is not necessarily so: by applying the connection-commutation technique to a Hall element working in the Hall voltage mode, one can strongly reduce offset and $1/f$ noise. An equivalent technique applicable in the Hall current mode and in the magnetoresistance mode does not exist (yet).

It has been seen that the current-related magnetic sensitivity of an extrinsic Hall plate is not very temperature-dependent [11]. On the other hand, both the current-deflection effect and the magnetoresistance effect depend directly on the mobility of charge carriers, which is strongly temperature-dependent. These two facts explain the predominance of the use of the Hall elements in the Hall voltage mode of operation over the other two modes.

6.2.2 Hall Sensors Versus Ferromagnetic Magnetoresistors

The only other solid-state magnetic field sensors with a practical importance comparable with that of Hall devices are the magnetic sensors based on the magnetoresistance effect in thin ferromagnetic films. We shall now briefly review the main properties of commercially available Hall magnetic sensors and of thin-film ferromagnetic magnetoresistors. Then we shall compare their characteristics and discuss their respective positions in the landscape of industrial applications.

6.2.3 Performance of Integrated Hall Magnetic Sensors

By way of example, we shall now briefly look at the characteristics of a few leading commercially available "Hall ASIC" (application-specific integrated circuit) single-axis magnetic field sensors. The first two (Hall-1 and Hall-2) are conventional integrated linear Hall sensors. They are made in BiCMOS and CMOS technology, respectively. The third one (IMC/Hall) is the CMOS-integrated Hall ASIC combined with integrated magnetic concentrators (IMC). The characteristics of these Hall sensors are summarized in Table 6.1.

We see that the integration of magnetic concentrators (IMC) brings about a higher magnetic sensitivity and lower equivalent offset and noise; this results in a better magnetic field resolution. The magnetic gain associated with IMC makes possible also the realization of a larger bandwidth. Recall also that conventional Hall magnetic sensors respond to a magnetic field normal to the chip surface, whereas an IMC/Hall sensor responds to a magnetic field parallel with the chip surface.

6.2.4 Performance of Ferromagnetic Magnetoresistors

The magnetoresistance effect in ferromagnetic materials was discovered almost 150 years ago, long before the Hall effect. Today this effect is called anisotropic magnetoresistance effect, and is used commercially in the form of magnetic field sensors called anisotropic magnetoresistors (AMRs). The typical magnetic

Table 6.1 Characteristics of some integrated Hall single axis magnetic sensors [12].

	Technology			
	BiCMOS	CMOS Sensor	IMC/CMOS	
Characteristics	Hall-1	Hall-2	IMC/Hall	Unit
Full-scale field range (FS)	±40	±14	7	mT
Sensitivity	50	140	300	mV/mT
Bandwidth (BW)	30	0.13	120	kHz
Linearity error within FS	0.1	1	0.5	%FS
Hysteresis error	0	0	20	μT
Equivalent offset	3	0.06	0.45	%FS
$1/f$ noise density at 1 Hz	0.3	—	0.3	μT/$\sqrt{\text{Hz}}$
White noise density	0.2	1.3	0.03	μT/$\sqrt{\text{Hz}}$

sensitivity is 10 mV/mT for a bias voltage of 1 V. This corresponds to a voltage-related sensitivity of about 10 V/VT, which is 10 times better than that of the best silicon Hall plate. Why is the magnetoresistance effect in a thin film so much stronger than the Hall effect in semiconductors? The explanation of this enigma is that in MRs, the high sensitivity is not only the merit of the appropriate galvanomagnetic effect. Instead, the origin of the high sensitivity is the result of a synergy of two distinct effects [13]: the one is, a strong magnetic flux concentration effect; and the other is a mediocre inherent magnetoresistive effect. Another much used MR has the structure of a sandwich of thin ferromagnetic and nonferromagnetic films. Its operation principle is based on the relatively recently discovered giant magnetoresistance effect (GMR). Paradoxically, a typical GMR sensor is less magnetic-sensitive than an AMR, but has much higher saturation field. There is a sizeable literature on AMRs and GMRs.

Table 6.2 lists the main characteristics of some commercially available general-purpose AMRs and GMR magnetic sensors. The full-scale measurement ranges of AMR and GMR sensors are limited by a magnetic saturation of the corresponding ferromagnetic films. Related with their structure is also the hysteresis phenomenon. A very unpleasant feature of AMR is the possible reversal of the internal magnetization of the magnetoresistor film (flipping), which causes a reversal of the sign of the output signal. A magnetic field shock (disturbing field) may demagnetize the film and so deteriorate the sensitivity of the sensor. These problems can be efficiently eliminated by magnetizing back and forth the film using an associated electromagnet. For GMR, a strong magnetic shock is a more serious problem: if a GMR is exposed to a strong-enough magnetic field (destroying field), it will be irreparably damaged.

Table 6.2 Characteristics of some AMR and GMR single-axis magnetic sensors [14].

Characteristics	AMR-1	AMR-2	GMR-1	Unit
Full-scale field range (FS)	0.6	0.6	7	mT
Sensitivity	50	64	27.5	mV/mT
Bandwidth (BW)	5	1	1	MHz
Linearity error within FS	1.4	6	2	%FS
Hysteresis error	0.3	0.5	3	%FS
Equivalent offset	30	19	10	%FS
$1/f$ noise density at 1 Hz	1	n.a.	16	$\mu T/\sqrt{Hz}$
White noise density	0.1	0.1	0.2	$\mu T/\sqrt{Hz}$

6.2.5 Integrated Hall Sensors Versus AMRs and GMRs

Let us now compare the characteristics and application areas of commercially available integrated Hall, AMR, and GMR magnetic sensors.

Hall plates have a relatively small magnetic sensitivity. This has two important consequences: first, detecting small magnetic fields is difficult; and second, a Hall sensor system may require a big electronic gain, which limits the achievable bandwidth of the sensor system. On the other hand, Hall sensors show no saturation effects at high magnetic fields; and, due to their compatibility with microelectronics technology, they can be integrated with interface electronics, which result in a low-cost smart magnetic sensor microsystem. Therefore, integrated Hall sensors are preferably used at higher magnetic fields. It was generally believed that Hall sensors are applicable only at magnetic fields well above 1 mT, and the measured magnetic field had to be perpendicular to the sensor chip.

AMR magnetic sensors have high resolution and high bandwidth, but they saturate at a rather small magnetic field (<1 mT) and they may require a complex resetting procedure. GMR magnetic sensors also have high resolution and high bandwidth, can operate also in the millitesla range, but have a high hysteresis and can be destroyed by a not-so-high magnetic field. Both AMR and GMR sensors respond to a magnetic field parallel with the MR layer.

Generally speaking, ferromagnetic MR sensors were considered as highly sensitive and good for small magnetic fields, whereas Hall sensors were considered as less sensitive and good only for high magnetic fields.

The concept of Hall magnetic field sensors based on an integrated combination of Hall elements and magnetic flux concentrators has been developed over the last years, and the first IMC Hall ASICs are now commercially available [12]. The IMC functions as a passive magnetic amplifier, boosting the performance of the Hall sensor. So the resolution and the bandwidth of an IMC Hall sensor approach those of ferromagnetic MR sensors. Moreover, seen from outside, an IMC Hall sensor responds to a magnetic field parallel with the chip surface, much as an MR sensor does. We conclude that AMRs and GMRs are still much better than modern Hall ASIC sensors only in the field of high-frequency AC magnetic measurements: MRs are preferred for both higher resolution and larger bandwidth. But at lower frequencies, the difference becomes smaller. For DC fields, an AMR or GMR are good only if some kind of chopper stabilization (switching, set/reset) is applied; otherwise, the resolution of the IMC Hall ASIC is much better than that of GMR and approaches the resolution of a nonswitched AMR. Therefore, the IMC Hall sensor ASICs are about to bridge the gap between AMR, GMR, and traditional silicon-integrated Hall magnetic sensors.

6.3 A Conclusion on NMR and ESR Spectroscopy

Nuclear Magnetic Resonance (NMR) spectroscopy and Electron Spin Resonance (ESR) spectroscopy are two widely used spectroscopic techniques to infer structure and properties of complex molecules. Both these methods use angular momentum, pure "spin" or "total angular momentum" of the relevant particles to extract molecular structural information. In ESR spectroscopy, molecules in a state containing unpaired electrons, i.e. with nonzero spin-angular momentum (molecules in nonsinglet states) are placed in constant magnetic field. For example, the ESR spectrum of unpaired electrons in transition metal complexes contains information on how ligands are arranged around the metal ion. Fine structure results from the interaction between electrons and nuclear spin and this can serve as a fingerprint for molecular, nuclear, and electron spin density. Likewise, in NMR spectroscopy, a composite system of nuclei in molecules with nonzero nuclear spins is placed in a constant magnetic field. Similarly, the NMR spectrum resulting from transitions between different states of nuclear spin system contains a wealth of information regarding chemical environment of the nuclei, and this is used to extract structural information of molecules. Despite their widespread use, the underlying principles of both these spectroscopic techniques are fairly simple. In fact, the machinery needed for a quantum mechanical description of the basic phenomena of these methods mainly involves the angular momentum theory. In addition, we would require basic first-order (and sometimes second-order) perturbation theory, and variational method and related secular problem. The Hamiltonians used are simple enough to enable manual solution of first-order perturbation theory and the secular equations.

6.3.1 Differences Between NMR and ESR

6.3.1.1 Resonant Frequency
One important difference between NMR [15] and ESR is that in ESR the resonant frequencies tend to be much higher, by virtue of the 659 times higher gyromagnetic ratio of an unpaired electron relative to a proton. For example, a typical magnetic field strength used in ESR spectrometers is 0.35 T, with a corresponding resonant frequency of about 9.8 GHz. This frequency range is known as "X-band," and the spectrometer as an "X-band ESR spectrometer." Such spectrometers are readily available "off the shelf" from a (small) number of commercial sources. X-band ESR spectrometers are typically used to study small solid samples or nonaqueous solutions up to a few hundred microliter in volume. They cannot be used for biological samples, or for in vivo studies, because of the strong nonresonant absorption of microwaves at 9.8 GHz. For that reason, ESR spectrometers (and imagers) have been constructed to operate at lower magnetic fields, and correspondingly lower frequencies, including at "L-band" (about 40 mT and 1 GHz) to study mice and "radiofrequency" (about 10 mT and 300 MHz) to study rats.

6.3.1.2 Relaxation Times
The second important difference between NMR and ESR is the typical relaxation times encountered. In biomedical proton NMR the relaxation times T1 and T2 are typically of the order of 0.1 to 1 seconds. In biomedical ESR the

equivalent electron relaxation times are a million times shorter, i.e. 0.1 to 1 μs! The extremely short relaxation times have important implications on the way in which ESR measurements are carried out.

6.3.1.3 Differences Between ESR and NMR Imaging

ESR imaging is very closely related to NMR-based techniques being the type of magnetic resonance used (i.e. ESR vs. NMR) and the method of signal detection. However, the idea at the heart of ESR imaging is exactly the same as the basic component of MRI, namely the use of a magnetic field gradient to "encode" spatial information about the sample into the magnetic resonance signals. Problems with ESR imaging can be concluded as follows: "Pure" ESR imaging is capable of imaging the distribution of many free radical contrast agents in vitro and in vivo. However, it does suffer from some problems:

1) ESR line widths tend to be large (another way of saying that ESR relaxation time is short), resulting in relatively poor spatial resolution in many cases – images can appear blurred in comparison with proton MRI.
2) The majority of biomedical ESR studies make use of free radicals with relaxation times that are too short to be detected by time-domain ESR. In this case it is necessary to use a radically different method of signal detection, namely continuous-wave (CW) ESR. Slice-selective excitation is not possible in CW-ESR imaging, so two-dimensional imaging with well-defined slices cannot be done. Imaging tends to be 3D, which either takes a long time, or else sacrifices spatial resolution.
3) ESR images show only the location of free radicals, so the anatomy cannot be visualized.

6.3.1.4 ESR Applications

The majority of applications of ESR to date have involved the use of stable free radical "probes". As well as these soluble probes, solid probes also play a role, in single crystal or particulate forms, especially in the measurement of oxygen concentration [16]. ESR is able to obtain useful information by virtue of the dependence of the ESR spectral characteristics (line width and/or line splitting) on physical and chemical parameters. This is the basis of "ESR oximetry" [17] and of pH measurement by ESR, using pH-sensitive nitroxides [18]. More challenging is the detection of naturally occurring free radicals, such as oxygen-derived radicals and nitric oxide; despite difficulties associated with low concentrations and short lifetimes of these molecules, they can be detected with the aid of chemical stabilization methods called spin-trapping [19]. Most biomedical applications of ESR to date have been in vitro, or in vivo in small-animal models, though at least one research group has begun to use ESR spectroscopy in a clinical context, for oximetric measurements [20].

6.4 Superconductive Quantum Interference Devices

Superconductive quantum interference devices (SQUIDs) are today's most sensitive devices for the detection of magnetic flux with energy resolutions approaching the quantum limit. They have a wide and flat frequency response ranging from DC to several GHz [21]. SQUIDs can be used as sensors for any physical quantity that can be transformed into magnetic flux, such as current, voltage, magnetization and susceptibility, displacement as well as temperature and others. They are therefore very versatile and can address a large variety of applications. To exploit the superior sensitivity of low-temperature superconductors (LTS) DC SQUIDs, however, a low operation temperature, of 4.2 K and below is mandatory. The need for cryogenics is a significant barrier to the widespread application of SQUIDs since both the operator's convenience and the system costs are impaired.

Fortunately, during the last years, general demand has advanced the development of cryocoolers which are now commercially available in a variety of models. However, to use these mechanical coolers, the measurement chamber is typically magnetically shielded, to attenuate magnetic and vibrational noise from the cryocooler. If these noise sources can be reduced considerably at reasonable expenses, a variety of potential markets may be opened. Current technology development is mainly towards a further decrease in junction capacitance and accordingly a downsizing of the Josephson junctions, as well as yield and parameter spread optimization. This will result in a strong sensitivity increase of SQUIDs. Moreover, small line-width devices enable the operation in ambient magnetic fields. Revealing the origin of magnetic flux noise may further improve the sensitivity of modern SQUIDs.

6.4.1 SQUID Fabrication Trend

Fabrication of planar, thin-film low-temperature SQUIDs using the Nb/Al – AlO$_X$/Nb processes is already a mature technology, in which only incremental progress can be expected, aimed mostly at optimizing yields and reducing the manufacturing costs [22]. An example is a possible wider introduction of less expensive lift-off processing, which can partially replace dry etching. The future of the technology is linked to the demand for most sensitive SQUID sensors and amplifiers. The former find use, for example, in some biomagnetic research and diagnostic applications, the latter are irreplaceable in scientific instrumentation and as detector readout devices. Despite the enormous progress that has been made over the last decade in terms of device performance, the numerous practical applications of high-temperature superconductors (HTS) SQUIDs and the availability of a few commercial products, there are still a number of problems that need to be solved to turn HTS fabrication into a mature technology.

6.4.2 Trends in SQUID Electronics

In the past decades, both rf and DC SQUID electronics have been improved considerably. Stimulated by the success of early HTS rf SQUIDs, practical techniques were developed to increase the bias frequency from the traditional values near 30 MHz up to ≥1 GHz [22]. Today, HTS rf SQUIDs with well-designed readout electronics can achieve noise levels comparable to those of their DC SQUID counterparts. For DC SQUIDs, the research and development activities concentrated on simplifying the readout electronics for multichannel applications and improving the dynamic performance for unshielded or high-frequency applications. Considerable increases in the slew rate and bandwidth were achieved by raising the modulation frequency from the traditional 100 kHz range up to some 10 MHz or by using direct readout without flux modulation. SQUIDs with cryogenic readout electronics or digital single-chip SQUIDs might achieve superior dynamic performance, but such devices are not yet ready for practical applications, representing a challenge and prospect for the future.

Many sets of SQUID electronics have been quite bulky. Particularly in the case of a multichannel system, the readout electronics should be as small as possible, and also should consume little electrical power. Nowadays, the use of surface-mounted electronic components makes possible a remarkable miniaturization, and – at the same time – can lead to a reduced power consumption and improved high-frequency behavior. The use of both digital circuits and surface-mounted components results in compact, fully automated, SQUID electronics which adjust the bias point of the SQUID even in a magnetically noisy environment without compromising the sensitivity.

6.4.3 Trends in SQUIDs for Nondestructive Evaluation of Materials

Numerous examples of the successful use of SQUIDs for nondestructive evaluation (NDE) have been presented. In almost all cases, SQUID testing was superior to other "conventional" NDE techniques [23]. Compared to conventional electromagnetic NDE devices, prototype SQUID systems have proved superior in performance in almost all cases. In some cases, such as the detection of inclusions in aircraft turbine discs, SQUIDs have been used commercially for routine inspections. In other cases, e.g. aircraft wheel hub inspection, the industrial partners have refrained from commercialization for economic reasons. Taking into account the cost of developing the prototype into a commercial product and the size of the likely market, the cost was estimated to exceed the expected profit. The same holds true for magnetic flux leakage inspection of steel structures on oil platforms: although the feasibility has been clearly demonstrated, the huge engineering effort required to make rugged a delicate SQUID system to the point at which it could be operated in the hostile environment of underwater structures was too elaborate to contemplate. In still other cases such as bridge inspection and magnetic particle detection, the SQUID competition led to progress in the development of NDE methodology.

Evidently, the most important future development requirement is user-friendliness. The technical user wishes to have a "turnkey" system where he does not have to be concerned with the complex superconducting and cryogenic technology inside it. Successful industrial SQUID uses of NDE are expected provided that (i) the focus is laid on applications with unique SQUID advantages where the larger system cost is acceptable, (ii) versatile signal interpretation tools are implemented, and (iii) improved SQUID reliability and handling are attained. SQUID NDE groups worldwide are concentrating on these issues.

References

1 Lenz, J. and Edelstein, S. (2006). Magnetic sensors and their applications. *IEEE Sensors Journal* 6 (3): 631–649.

2 Zuo, S., Fan, H., Nazarpour, K., and Heidari, H. (2019). A CMOS analog front-end for tunnelling magnetoresistive spintronic sensing systems. In: *2019 IEEE International Symposium on Circuits and Systems (ISCAS)*, 1–5. IEEE.

3 Nabaei, V., Chandrawati, R., and Heidari, H. (2018). Magnetic biosensors: modelling and simulation. *Biosensors and Bioelectronics* 103: 69–86.

4 Zuo, S., Nazarpour, K., and Heidari, H. (2018). Device modeling of MgO-barrier tunneling magnetoresistors for hybrid spintronic-CMOS. *IEEE Electron Device Letters* 39 (11): 1784–1787.

5 Li, H., Shrestha, A., Heidari, H. et al. (2019). Magnetic and radar sensing for multimodal remote health monitoring. *IEEE Sensors Journal* 19 (20): 8979–8989.

6 Heidari, H., Bonizzoni, E., Gatti, U. et al. (2016). CMOS vertical Hall magnetic sensors on flexible substrate. *IEEE Sensors Journal* 16 (24): 8736–8743.

7 Heidari, H., Bonizzoni, E., Gatti, U., and Maloberti, F. (2015). A CMOS current-mode magnetic Hall sensor with integrated front-end. *IEEE Transactions on Circuits and Systems I: Regular Papers* 62 (5): 1270–1278.

8 Park, J.-H., Vescovo, E., Kim, H.-J. et al. (1998). Direct evidence for a half-metallic ferromagnet. *Nature* 392 (6678): 794.

9 Popovic, R., Flanagan, J., and Besse, P. (1996). The future of magnetic sensors. *Sensors and Actuators A: Physical* 56 (1-2): 39–55.

10 Boero, G., Demierre, M., and Popovic, R. (2003). Micro-Hall devices: performance, technologies and applications. *Sensors and Actuators A: Physical* 106 (1–3): 314–320.

11 Popovic, R.S. (2003). *Hall Effect Devices: Magnetic Sensors and Characterization of Semiconductors*. CRC Press.

12 Popovic, R. and Schott, C. (2002). Hall ASICs with integrated magnetic concentrators. *Proceedings of the Sensors Expo and Conference*, Boston, USA (23 September 2002), pp. 23–26.

13 O'Handley, R.C. (2000). *Modern Magnetic Materials: Principles and Applications*. Wiley.

14 Popovic, R.S., Drljaca, P.M., and Schott, C. (2002). Bridging the gap between AMR, GMR, and Hall magnetic sensors. In: *2002, 23rd International Conference on Microelectronics: MIEL 2002. Proceedings* (Cat. No. 02TH8595) (12 May 2002) vol. 1, 55–58. IEEE.

15 Lei, K.-M., Heidari, H., Mak, P.-I. et al. (2016). A handheld high-sensitivity micro-NMR CMOS platform with B-field stabilization for multi-type biological/chemical assays. *IEEE Journal of Solid-State Circuits* 52 (1): 284–297.

16 Dunn, J.F. and Swartz, H.M. (2003). In vivo electron paramagnetic resonance oximetry with particulate materials. *Methods* 30 (2): 159–166.

17 Swartz, H.M. and Dunn, J.F. (2003). Measurements of oxygen in tissues: overview and perspectives on methods. In: *Oxygen Transport to Tissue XXIV. Advances in Experimental Medicine and Biology*, Vol. 530 (ed. J.F. Dunn J.F. and H.M. Swartz), 1–12. Boston, MA: Springer.

18 Khramtsov, V.V., Grigor'ev, I.A., Foster, M.A., and Lurie, D.J. (2004). In vitro and in vivo measurement of pH and thiols by EPR-based techniques. *Antioxidants and Redox Signaling* 6 (3): 667–676.

19 Berliner, L.J., Khramtsov, V., Fujii, H., and Clanton, T.L. (2001). Unique in vivo applications of spin traps. *Free Radical Biology and Medicine* 30 (5): 489–499.

20 Salikhov, I., Walczak, T., Lesniewski, P. et al. (2005). EPR spectrometer for clinical applications. *Magnetic Resonance in Medicine: An Official Journal of the International Society for Magnetic Resonance in Medicine* 54 (5): 1317–1320.

21 Grosz, A., Haji-Sheikh, M.J., and Mukhopadhyay, S.C. (2017). *High Sensitivity Magnetometers*. Springer.

22 Clarke, J. and Braginski, A.I. (2004). *The SQUID Handbook, Vol. 1, Fundamentals and Technology of SQUID and SQUID Systems*. Wiley-VCH.

23 Clarke, J. and Braginski, A.I. (2006). *The SQUID Handbook: Applications of SQUIDs and SQUID Systems*. Wiley.

Index

Magnetic Sensors for Biomedical Applications, First Edition. Hadi Heidari and Vahid Nabaei.
© 2020 by The Institute of Electrical and Electronics Engineers, Inc.
Published 2020 by John Wiley & Sons, Inc.

IEEE Press Series on Sensors

Series Editor: Vladimir Lumelsky, Professor Emeritus, Mechanical Engineering, University of Wisconsin-Madison

Sensing phenomena and sensing technology is perhaps the most common thread that connects just about all areas of technology, as well as technology with medical and biological sciences. Until the year 2000, IEEE had no journal or transactions or a society or council devoted to the topic of sensors. It is thus no surprise that the IEEE Sensors Journal launched by the newly-minted IEEE Sensors Council in 2000 (with this Series Editor as founding Editor-in-Chief) turned out to be so successful, both in quantity (from 460 to 10,000 pages a year in the span 2001–2016) and quality (today one of the very top in the field). The very existence of the Journal, its owner, IEEE Sensors Council, and its flagship IEEE SENSORS Conference, have stimulated research efforts in the sensing field around the world. The same philosophy that made this happen is brought to bear with the book series.

Magnetic Sensors for Biomedical Applications
Hadi Heidari, Vahid Nabaei